Footsteps in Science

J.H.U. BROWN, PH.D.
Professor
University of Houston
Houston, Texas

UNIVERSITY PRESS OF AMERICA

Lanham • New York • London

University Press of America®, Inc.
4720 Boston Way
Lanham, Maryland 20706

3 Henrietta Street
London WC2E 8LU England

Printed in the United States of America
British Cataloging in Publication Information Available

Grateful acknowledgement is given for permission to reprint material from the following copyrighted sources:

The High Cost of Healing © 1985,
Human Science Press, New York, NY

Telecommunication in Health Care © 1982,
CRC Press, Boca Raton, FL

Educating for Excellence © 1989,
Greenwood Group Publ., Westport, CT

History of Biomedical Engineering, Special Edition of
Engineering in Biology and Medicine © 1991,
IEEE Press, New York, NY

Library of Congress Cataloging-in-Publication Data

Brown, J.H.U. (Jack Harold Upton), 1918–
Footsteps in science / J.H.U. Brown.
p. cm.
1. Science. 2. Scientists. 3. Research.
4. Brown, J.H.U. (Jack Harold Upton), 1918– .
5. Scientists—United States—Biography. I. Title.
Q172.B76 1992 500—dc20 92–10033 CIP

ISBN 0–8191–8693–7 (cloth : alk. paper)
ISBN 0–8191–8694–5 (pbk. : alk. paper)

The paper used in this publication meets the minimum requirements of American National Standard for Information Sciences—Permanence of Paper for Printed Library Materials, ANSI Z39.48–1984.

TABLE OF CONTENTS

PREFACE

I am not presumptuous enough to write an autobiography. Nor is this book intended to reflect totally my life and the events in it. Rather it is a demonstration of the way science and scientists develop. Thomas Kuhn in his remarkable book "Structure of Scientific Revolutions" has pointed out that science does not arise *de novo* but occurs as a result of accretions of ideas, often from long ago. This book demonstrates such accumulations in one person's life. There may be apparent repetitions of information but the development of a science requires that repetition. By the same token, the areas of science have changed over the years as ideas developed and new congregations of facts emerged. I have tried to show such changes as they occurred.

In most cases real names have been used. I apologize if I have misused someone or misreported a fact. I hope the sins are those of omission rather than commission. I have been fortunate to have lived in a time when science was rapidly advancing and I was fortunate to hold positions where I was often in the forefront of that development. The events would not have happened without the support of many individuals who guided me along the way. It is impossible to name all of them here but I must single out Frederick Stone who you will meet in these pages.

The events related are not chronological. I have chosen to illustrate the development of ideas rather than to follow the day to day experience of one scientist which can often be boring and routine. Science is a unitary process. As a result some chapters overlap into others. This is deliberate. I have attempted to cross reference between chapters and sections so that the reader can intermix events of scientific endeavor.

The book is divided somewhat arbitrarily into two sections. The first section is a description of how a particular facet of an area of science was developed and the results which emerged. The second section is a partial autobiography to fit the events of my life into the experimental development outlined in the first section. The book could have been arranged in the reverse order, but I believe the reader will be more attracted to the scientific aspects.

I have included references at the end of many chapters In most cases they are my own and serve to illustrate the points made in the chapter. In no case are they definitive. They provide additional information to pursue an idea further and for this reason they are not indexed within the chapters or in a bibliography.

J.H.U. Brown

PART I

FOOTSTEPS IN SCIENCE

CHAPTER 1

INTRODUCTION

There are several types of behavior in scientists. One is the person who conceives of an idea, develops it rapidly and reaches a conclusion and then drops it for something else. The second is the person who exploits an idea to the last grain of truth. Science requires both kinds. Someone must erect a framework of bare bones and someone must fill in the framework with bricks and mortar. The latter type of working is foreign to me because I rapidly become bored with repetition and with the last details of a project. Solomon Snyder, the well known neurobiologist has said that he prefers to do experiments that can be completed in one day. This permits him to do ten different experiments in ten days, rather than to exploit the ideas of one experiment for all ten days. In addition, if the experiment is not a success, only one day has been wasted. I fully agree.

Scientists are also either pragmatic or theoretical. I have been listed in the former category. The pragmatic scientist looks for the practical use of his results. Science for science's sake have little place in his vocabulary. The pragmatist has no objection to "basic"research but a goal to be reached is always in mind. The theoretical scientist is the opposite. He does research for research's sake. The results are sufficient unto themselves and if someone can later find a use for them, so much the better. It is this type of scientist who is seldom cited (Chapter 8) or who spends a lifetime on investigating the left hind leg of the flea. He may become a world authority in a narrow field, but no one else cares.

In another classification, scientists can be gathered into groups of lumpers or splitters. The splitter is determined to find the final constituent of a problem. The splitter physicist uses the cyclotron or supercollider to split the atom into smaller and smaller particles in the hopes of finding the ultimate building blocks of nature. The biologist is interested in breaking down the structure of biological matter into amino acids, enzymes, or ultimate cell structure. The lumper biologist, on the other hand, is interested in how the organism works as a whole. True he must understand something of the internal workings of the cell to understand how billions of cells interact to form systems and how those systems are controlled. But the understanding of the minutia is only sufficient for the moment to enable the synthesis of ideas. The contrast is clearly shown in astronomy. Astronomers look in two directions. One group attempts to understand the "big bang" and how the universe is formed and develops. A second group is interested in quantum mechanics which is physics of the infinitesimal. Einstein spent a large part of his life attempting to combine the two into a unified theory. No one has done so and many believe that the task is impossible because the two theories are based on incompatible approaches. The lumper and splitter in biology are in much the same situation. I am a lumper interested in how systems perform. To date there is no biological synthesis which explains how amino acids and other molecules are organized into the complete organism with reproducing cells and organelles.

The pragmatic biologist is interested in the way science can be applied in daily life, in the relationship between the basic and applied sciences. He accepts the data as it is measured and attempts to make use of it. The engineer is the ultimate pragmatist. It is his responsibility to make something workable of many components parts. A few years ago the engineering schools decided to scrap practical teaching in engineering in favor of "engineering science". They were rapidly forced to revise their ideas when companies began to complain that the young engineers were knowledgeable but could not "do" anything.

For these reasons I have been called a dilettante in science. It is true that I have been a biomedical engineer and an endocrinologist, an environmentalist and a medical administrator, an iconoclast regarding medical education and an author on a wide variety of topics. There is a reason.

I have been fortunate to be in the beginning of several projects which will described below. The creation of the Papago health project, development of the minicomputer, creation of a cancer treatment drug, new forms of medical records, and the origin of Bioengineering were all projects in which I had a hand at starting but not in developing to the fullest. Others were involved in that process. Without arguing whether my teaching is good or bad, I have been a dilettante there as well as an instructor in physics and in physical chemistry at the college level, professor of physiology and of business, professor of biology and public administration and an adjunct professor of community medicine and art .

There are no excuses for my life in science. It has been most enjoyable. One hears arguments that it is the lifelong pursuit of an idea and all of its ramifications which results in the most reward. That is not totally true. I have been fortunate in election to several major scientific organizations as Fellow and to the National Academy of Engineering. It has been luck in being in the right place, partly in knowing persons willing to help at the right time, and luck in picking projects which had not been previously attempted and which were successful.

Another major attribute of the successful scientist is the ability to see a field of science developing and take advantage of it. This requires fast feet and an ability to change directions. Nothing is sadder than to see a scientist (?) working on the same problem developed for his Ph.D. dissertation twenty years ago.

A good scientist must be both a visionary and a technician. The visionary must have a degree of serendipity in order to be able to relate what may seem disparate facts into a cohesive whole. The ability to have concepts outside the realm of daily tasks is critical. Otherwise, the scientist is just a technician, performing meticulous tests, amassing data, and assembling statistical information. But unless both are combined, neither will be a success.

Success in science is coupled with success as a scientific politician. Election to office in national organizations, promotion at a university, and the advertising of one's ability requires political acumen. In addition, the present day atmosphere of negative attitudes toward science requires that the scientist enter the political arena if only to preserve grants and university support.

One cannot be a successful scientist unless one is able to write well. The ability to take data in its raw form and convert it to simple prose is critical. Rarer but of equal importance is the person who can take data from many sources and convert it into a book that explains a phenomena or provides a philosophy of science. The talent to write a book, an article, or a news item, all contribute to the ability to present ideas and have them accepted. They serve to provide ammunition for promotion and pay and to bring ideas and personalities before the public.

In succeeding chapters we shall look at rapid changes of direction and forces of change and where they have lead one scientist. The reader must be the judge about whether the process was worthwhile.

CHAPTER 2

FROM IDEAS TO IMPLEMENTATION

There is no greater satisfaction in science than to have an idea, see it worked out in experimentation and finally have it brought to fruition in a practical device. Most scientists are interested in devices. But the interest is largely confined to their use. Computers are the greatest of all gadgets; my interest in computers began when I became the Assistant Director for Operations of the Division of Research Resources (DRR) of the National Institutes of Health (NIH) in 1963. The division had inherited from other institutes at NIH a program of computer development related mostly to computation and use of computers in spectral analysis, and other large scale computational processes requiring large memories-the number crunchers. We became interested in Bioengineering and further development of computers (Chapter 3). The opportunity came when we were approached with a proposal by Jerry Cox and Wes Clark from the Lincoln Labs associated with MIT to develop a small computer for individual use in the laboratory-a minicomputer. It was a good idea. In collaboration with Orr Reynolds of NASA we appropriated about $1,200,000 between NASA and the NIH to enable us to launch the project. Smaller sums were contributed from other institutes at NIH.

An old garment factory in the rear of the MIT campus was available and the second floor was rented and converted to computer manufacture. Over the next many months the team developed a computer using the ferritecore memory invented by a member of the team, Bill Papian, a small tape drive devised by Wes Clark, and other innovations to create the LINC-8. The experimental team of Clark, Molnar, Cox, and several others designed the machine and purchased parts. I served as the NIH representative manager and traveled from NIH to Boston frequently. The name, LINC was derived from two sources, Lincoln Labs where most of the team originated, and the idea of a machine to link laboratory activities. Later the name was changed to Laboratory Instrumentation Computer.

A major problem was how to test a new machine with no specific purpose except to be used in the laboratory for computation as necessary and at the same time to demonstrate its usefulness in a variety of situations. We decided to ask for applications from the scientific community. From a large number that were submitted, we selected fifteen of the most promising representing a wide range of potential use from online monitoring of nerve transmission to statistical analysis. The successful applicants were brought to MIT to assemble their own machines, learn the programming, and develop a feel for troubleshooting. The machines were then sent to the individual labs for a year of testing in laboratory conditions. They were an unqualified success. For the first time an individual scientist had at his disposal in his (no females were included) laboratory, a powerful computing machine. The machine was small by the standards of computers of that day that occupied suites and required enormous airconditioning equipment. True, the machine was large by present day standards. Wes Clark had specified that it could not be higher than a man and must occupy a small floor space. The finished computer was five feet high and two by three feet in floor area. It had a small memory by present day standards- many hand held computers now have an equal capacity. In addition, it was hand assembled from many parts; today the same computer can be mass produced by the millions using microchips.

There had been trouble along the way. MIT had appointed Murray Eden and Walter Rosenblith as the representatives of the university to correspond with the government representatives. It shortly became evident that the system could not work- Eden and Rosenblith assumed the role of dictating actions of the team, controlling the methods of review and assignment of machines, and in general making the experimental team very unhappy. Dr. Fred Stone (then Director of DRR) and I went to MIT, met with Jerome Weisner, the Provost, and told him of our concerns. He was unsympathetic and essentially took the position that "MIT would run the show."

We were fortunate to have at the time on our Advisory Council Dr. Edward Dempsey, Dean of the Medical school of Washington University in St. Louis. Dr. Dempsey heard of our plight and approached his President, Dr. George Pake, who in turn offered to take the entire team to Washington University. Dr. Jerry Cox who worked on the LINC project was also a professor in the Institute for the Deaf at Washington University and his help was essential. In short order most the group moved and Jerry Cox became director of the new biomedical engineering institute. It is still in operation.

But we had successfully built and tested the first minicomputer. In the end we offered the plans to computer manufacturers to make LINC-8's. DEC (now Digital) was interested and Ken Olsen built the PDP-4 as a result. IBM showed no interest and lagged behind in minicomputer development for many years.

Afterward I moved to the Health Service and Mental Health Administration (HSMHA) in 1970 as Associate Deputy Administrator for Research. We were interested in research and delivery of services in emergency care, cost of health care and other aspects by establishing Emergency Medical Service Centers, HMO's, Experimental Health Delivery Systems and other clinically oriented programs, and it rapidly became apparent to me that one of the major problems in the health care system from the standpoint of both the delivery and the cost of care was the ability to handle the patient record and tie it to costs, charges, and care. The record was bulky, arranged chronologically, and a physician had to thumb through pages of irrelevant information to find the one piece of data needed for a diagnosis. It was then necessary for the information to be transferred to the auditor who matched, often unsuccessfully, the treatment provided with the charges to the patients. It occurred to me that the obvious solution was to have the patient carry his record with him/her in moving through the system so that it would be available and could be updated at each event. The costs could be directly tied to the treatment at the time it occurred. However, there was no suitable method. Microfiche had been proposed but this required expensive photography and much time to update the information. Magnetic strips on cards were available but were very low in carrying capacity. No other method existed.

Any new form of patient record had to accomplish several goals. It had to be non-erasable so that it was permanent, it had to contain all of the medical record, it had to be capable of updating at each encounter with the system, and it had to be inexpensive enough to be used by patients, physicians and hospitals and it had to be coded so that no unauthorized person could gain access.

In 1974 I moved to San Antonio as Director of the Southwest Research Consortium. There was a strong base of both applied and basic research within the organization and I tried to interest several groups in the idea of a new patient record. A break came on meeting Dr. C.B. Jackson, a practicing physician interested in computers. He was tall lanky Texan with a habit of peering over half frame glasses. He had worked closely with NASA and was an innovator of the first rank. In addition he was working with a commission of the American Medical Association attempting to classify diseases into a pattern that could be used for diagnosis and billing, the ICD (International Classification of Disease), which would be ideal for computerization of records. Jackson wanted to supply a computer based record system for the physicians in town and we discussed increasing the capacity of the major information mechanism then available. This was the magnetic strip. But nothing came of the idea. We were unable to achieve our goals, but we gained valuable information on processes which would not work.

In 1978, I moved to the University of Houston as Associate Provost for Research. Here I approached the College of Engineering and the College of Technology but aroused little interest. I talked to Frank Webber, then head of Family Practice at the University of Texas Medical Center and to Grif Ross, former director of NIH hospitals and now Dean of Medicine at Texas Medical Center in Houston and both were enthusiastic about the idea. Unfortunately both died within a short time of each other.

In the meantime I had gotten in touch with Bull-Schlumberger of France who had developed the idea of placing a computer chip on a credit card. The chip could contain perhaps 10 kb (kilobytes) of data and held more information than a magnetic strip and furthermore, it could be manipulated much more easily. We tried a chip for a short period but rapidly concluded that it was not feasible for our purposes as it could not contain a complete medical record and there was some question about security.

I had read in *Spectrum* about laser techniques and the 3M production of optical discs (CDs) using lasers for recording and reading high fidelity sound. The system appeared to be ideal for medical records. Dots were impressed by a laser beam on a matrix and were read by scanning the dots with the same beam. The system was ideal in that it could not be changed once entered, it was simple and almost fool proof, large quantities of information could be stored in a small space, and the information could be handled much like a floppy disc on a small computer. If the system could be placed on a credit card or similar device which could be carried by the patient, the possibilities were great.

I wrote to 3M asking about their techniques and suggesting that we might be able to use the optical method on a credit card to be carried on the person rather than on a revolving disc which would occupy too much room. We corresponded for some time but each individual which I contacted referred me to another and finally the whole thing petered out. I also made the same proposition to Beckman Instruments except that I proposed that the card be used to take analog data directly from the SMA machines then used in hospital analytical laboratories and place it in the patient record. Beckman was interested and paid my way to spend two days in Tarrytown, NY explaining my ideas. We looked at the output of Beckman machines and discussed recording and analysis of the data. At the end they thought that it would require as much as one billion dollars to retrofit and equip machines with the device and this was not practical from their viewpoint.

Knowing the Japanese technical ability with computers, I wrote to JETRO (Japanese External Trade Organization) and MITI (Ministry for International Trade and Industry) the Japanese organs interested in industrial/university/government cooperation outlining my ideas and asking if there was any interest. Within a short time K. Nishimura got in touch with me. Ken was a vice president of Sumitomo, a very large Japanese trading company. They had connections with CSK, a software company and had the ability to investigate laser card technique. Cards were purchased from Drexler in California which had developed a matrix on a credit card that could accept laser beam input and the consortium developed a reader-writer for the card along with the software to permit the card to be driven by a desk top computer. We were in business.

Ken and I realized that we must find a clinical test bed for the laser card. The medical connection was essential in order to test the card in real time for medical personnel and patients. Software had to be written in a format useful to the physician and other hospital personnel, data had to be assembled in a form from which it could be aggregated and combined to form diagnostic patterns useful in medicine. I had warned Ken that the programming would be difficult and expensive and much different from that usually attempted by CSK but he did not seem to be greatly concerned. We debated testing the use of the card in Houston and San Antonio. I had connections with Bexar Country Hospital in San Antonio and I proposed the system to them. They were not interested. I then began talks with two groups in Houston. Frank Webber, mentioned above, was interested, and Carlos Vallbona, Head of Community Medicine at Baylor College of Medicine was very interested.

Carlos had experience in working with William Spencer at Texas Institute for Rehabilitation Research (T!RR), one of the first investigators in computerized patient records and had developed his own and very good system of patient records. His computer experience plus his connection with the Harris Country Hospital District (HCHD)where he supervised the neighborhood clinics made his choice a good one as it provided a clinical test bed and an experienced computer user at the same time.

During this time, I had talks with the Surgeon General of the United States about using the credit card as a dogtag for military personnel to contain both medical and military records. The talks meandered along and I was finally put in touch with a young Naval officer at the Department of the Navy who had been given the task to determining how the new dogtag would be structured. We corresponded for many months but nothing happened. The Department of Defense (DoD) is still trying to determine the proper format for dogtags.

We now turned to Carlos Vallbona who was still very interested. As he was an employee of Baylor College of Medicine, any grant would have to come through the offices there and Sumitomo was very anxious to have an academic connection. I was made an adjunct Professor of Community Medicine and signed a contract with Sumitomo to develop the system. CSK assigned personnel to write the software and we were placed in contact with Junji Shoda, a young engineer at CSK, who came to Houston regularly to discuss developments. Sumitomo signed a contract with Baylor naming Vallbona and me as assignees with royalties to be paid when the project was in operation.

Software development was a major task. We had to take the Japanese idea of computer arrangement and in it place our ideas of software development. It required almost two years of development, redevelopment, and rewriting to obtain a satisfactory product which was useful to physicians.

The new system solved many problems. The acquisition and maintenance of medical records has long been a problem. The old hand written record has given way to computerized records in many cases, but there are still major difficulties with the system. If a system goes down, no records are available and in an emergency situation, valuable time is lost. Records cannot be transmitted from hospital, nursing home, hospice, emergency room or ambulatory clinic to other locations without interconnections which are difficult to maintain and expensive to construct. The obvious solution is for the patient to carry his own record and present it at each encounter with the system. In order to be satisfactory the record must be complete, up-to-date, and capable of being updated at each encounter. We successfully solved the problem. The laser card offered the possibility to have a complete record available at each encounter with the health care system as it can be carried by the patient and updated when the patient contacts the health care system..

The software is the critical element in the total system. It must fulfill several conditions. It must be accurate, unerasable, and readily accessible by clerks and physicians. It must be in a format which the physician will use. Summary data must be presented so that decision trees can be worked out in order to reach a diagnosis.

In addition, the physician must be free to add text information as he desires to supplement the coded data. The patient must be educated to carry the card at all times, he/she must understand that submission of the card at a treatment point gives permission to the health care system to access the data and must understand the card and how to handle it. All of these problems were addressed in the final system.

Data are placed on a standard sized credit card with a milliwatt laser beam that places 5 micron dimples on a special xurface . The card can be written upon using a standard format compatible with other systems. The record is divided into folders which can be accessed at will. The folders are grouped into sets to obtain data on several parameters as desired. In addition to the medical data, any other desired information can be entered on the card. The question of privacy of the patient is also solved because the patient carries the card with him and when it is tendered to the medical service unit, permission is given to access the data. In addition to the protection offered by a coding system, there is another and more critical advantage. The card cannot be corrected. Although old data can be deleted by software, it cannot be removed, so that any record is permanent and this is critical for patient and physician protection. Any data can be obtained by a single key stroke. Embedded in the program are adjuncts to aid the physician. The ICD (International Code of Diseases) codes are listed and the operator can go from code to disease entity or vice versa. Laboratory tests are entered in the system so that any test can be selected and normal values are compared to the patient data directly on the screen. With the development of the system, it became necessary to evaluate its operation in actual clinical trials. The evaluation involved several elements: card durability, patient acceptance, administrative use and costs, physician use, and the ability to make reliable and rapid diagnoses.

Sumitomo was interested in copyright and patents and therefore pressed for a name for the system which could be registered. We had a long discussion and I believe that Dr. Vallbona came up with the final name--"The Health Passport" (HPS. That name has been adopted by Sumitomo.

In the meantime, I wrote an article for a medical journal blaming the medical school training for much of the cost of medical care. The President of the Baylor College of Medicine took umbrage. It was suggested that I resign as professor. An arrangement was made that I would be a consultant to Sumitomo and be paid directly by them.

In 1987, I got in touch with the Director of the Veterans Hospital in Houston, John Sheehan, who appeared interested in the idea. His new director of computer services, Mike Bible, was also interested. We made a presentation of the system to the medical staff and received some support. About the same time I talked to Dr. Michael LeMaistre, Director of the M.D. Anderson Cancer Hospital in Houston who asked his people to look into the matter. There were a series of meetings with the medical staff but nothing happened. The VA project continued off and on for several years and is still viable.

In the meantime, Sumitomo/CSK was having troubles of its own. Several Japanese companies were interested in the laser technique but had difficulty in producing hardware which would read and write without error. The problem was finally solved by using three laser beams, one to read and write and two to define the pathway in which the active beam moved. The cards from Drexler were originally of poor quality and it became necessary to examine each one to determine if the quality was high enough to be used. This ran up the costs of the cards. Again the problem was solved by Japanese intervention in Drexler to assure a better quality.

In order to evaluate the system cards were issued to some patients who were compared to other patients in the clinic who had only paper records. Although the pa tients were directed to handle the card exactly as they would any other credit card, there was no instance where the card could not be read-the hardware worked perfectly. Patients indicated they strongly approved of the card system and would like to continue it as it saved time in the clinic. Ninety seven percent of all patients returned with the card at each visit.

Administrative personnel reported the card was easy to use, records could be located more rapidly, and the card appeared to save time and money. In order to evaluate the efficiency and effectiveness of the optical card system, "peer review" studies were conducted to determine the statistical difference in quality and efficiency of patient treatment when the HPS was compared to the usual clinic record procedures. Physicians were asked to perform simulated examinations of twenty patients. Each physician was given HCHD charts and HPS cards or printouts and received thirty minutes of instruction to familiarize him with the p.c. (personal computer) as only one had previous computer experience. They were asked to update active problems, devise a management plan including tests and procedures and prescribe drugs or medications. Each procedure was timed. Three senior physicians on the faculty of the Baylor College of Medicine were asked to create a system of evaluation of responses rating the responses as satisfactory or unsatisfactory in terms of correct diagnosis, correct treatment, and correct drug orders.

The physicians, who were previously unfamiliar with any computer, learned to use the computer and the system rapidly. A typical learning curve was obtained which flattened after 10 trials in about 25 minutes. On the average it took physicians well acquainted with the Harris County Hospital District system about 13.5 minutes to find vital information in the written record. Despite lack of familiarity with the computer, the same information was obtained from the card in about 10 minutes. In addition, the information obtained from the cards was more accurate and complete than that obtained from the hospital written record as determined by the examining board.

We finally launched a test program at Harris County Hospital District with about 100 patients. It was satisfactory, but evaluation revealed that the physicians had reservations about the way the data was presented and summarized which required additional work for programmers.

As the system became more and more useful we published a series of papers by invitation, some 30 in all in various journals and began to receive invitations to speak at international meetings. In two years I spoke by invitation at some 12 meetings and Dr. Vallbona spoke at several more. The International Medical Records Institute in Boston helped by announcing we had the only workable system. We also appeared on Voice of America and in articles in several popular magazines. Inquiries arrived from many countries and from all over the United States. Visits have been received from the U.S. Army, Argentina, Columbia, Mexico, Canada, many group practices in some 20 states, from individual physicians and from several countries in Europe. The government of Canada asked for information for their health care system. However, Sumitomo was reluctant to expand operations without standards for cards and writer/readers which could be accepted world wide. They therefore elected to confine their support to Texas.

We started a program with the Texas Rehabilitation Commission to provide records for children with disabilities, with the Veterans Hospital in Houston to maximize record availability in the clinics, and with a obstetrical practice at the Woodlands Hospital. NASA installed two reader/writers to test the system for keeping astronaut records. There was great interest.

There is now competition. For the first few years we had the only workable system but at present several Japanese firms (Canon, Olympus,etc.), Drexler, Optical Recording Corp. of Canada, and perhaps others have seen the potential and entered the field. In addition, Great Britain and France are pushing a medical record card using a computer chip.

The system is slowly expanding into new directions by investigating the use of the card as a student record system containing grades, degree plans, debit features for all costs, medical and other academic records and as a storage device for archives and library functions. A plan for handling medical records which would permit easy transfer of the elderly from hospital to nursing home to hospice to emergency room to home care with drug control and billing has been proposed. Other projects are on the horizon.

From a small beginning, the laser card record system is slowly expanding into a major industry. The uses of the system are unlimited. This is a long way from a large computer (although mini in concept) with 64 kb of memory to a credit card with a desk top computer handling 2.5 mb (megabytes) of medical records. The major advance was the development of a new computer memory, the laser card, which provided the possibility of a patient entering any part of the health care system and being able to present his records from all past encounters so that immediate and correct treatment can be given. This may be the future in medicine.

REFERENCES

Brown, J.H.U. *Developing a Health Care System.* Annals of Biomedical Engineering. 3: 3 (1975).

Brown, J.H.U., *The Optical Card in Medical Practice*. Computers in Biology and Medicine, 19: 375 (1989)

CHAPTER 3

THE BIRTH OF A NEW SCIENCE

It is not often that one can not only see the birth of a new form of science but actually participate in its generation. Biomedicalengineering can be defined as the applications of engineering and hard sciences theory and practice to the solution of biological problems. It came to prominence within the last thirty years although forms (bioelectric phenomena, for example) had been known for many years. However, the formulation of a definition and criteria for the discipline is of recent origin.

Like so many others, I came into biomedical engineering by the back door. Most physiologists are not unacquainted with Bioengineering principles and I was a physiologist. My real interest began when I came to NIH.

The activities of engineering in biology and medicine were scattered throughout NIH at that time. When I moved to NIH in 1960 (Chapter 14) a group had been organized in ultrasound research with Herman Schwan and John Jacobs with contributions from Otto Schmitt. NIH had a consultant in the field, Sam Talbot, with whom I traveled a great deal in the first year. His interests were largely in engineering. We were able to establish some program projects but most were in computer science.

When I became the Executive Secretary of the Physiology Training Committee of NIH in 1960, biomedical engineering as such did not exist. Fred Stone and I talked seriously about the relation of the physical sciences and the life sciences. The conclusion was that something should be done. Many possibilities were explored and within a few months of my arrival at NIH we organized a meeting of persons who might have an interest in the field. In December of 1961 a conference of 23 persons was held at NIH with Otto Schmidt as chairman.

The meeting was held in T-6, a temporary building left over from the war that stood where Building 31, the research administration building, stands today. It was a long, narrow room that barely held the table and chairs and it was hot, stuffy, and smoke filled (this was long before the days of smoking bans). I cannot remember all of the names but those of Herman Schwann, Otto Schmidt, John Jacobs, Art Guyton, Fred Grodins, and others come to mind.

We argued for a day about what biomedical engineering was, what the components were, and how research and training should be organized. We reached no agreement at that meeting or for many meetings over the following years. A combined training committee for physiology and biomedical engineering was organized in DGMS (Division of General Medical Sciences), and rapidly became a hot bed of argument. One side was ably defended by Art Guyton, a noted physiologist who argued that biomedical engineering was simply physiology in which the life sciences were better served by application of physical principles and the other side was equally vehemently defended by John Jacobs, a well known engineer who argued that a good engineer could learn all the biology he needed in no time and that engineering was an applied science but not necessarily applied to physiology alone. There were all shades of difference between the extremes.

Because of the differences, the joint committee did not succeed. The physiologists voted against all programs that did not include basic science as the fundamentals and the engineers voted against all programs that did not have a firm foundation in engineering. Site visits to potential programs became a source of debate whether programs should be in engineering schools or medical schools, whether and how much biology and engineering should be required and if programs were "too basic" or "too applied".

The science also suffered from the lack of identification. The English had a program called Clinical Engineering that they considered to be Bioengineering, but the biologist and the physiologists in this country objected to the "clinical" definition. The biologists objected to the term "Medical" although some proponents (Cesar Caceres), a physician, suggested "Medical Engineering" as the proper name for the new program. One night in Memphis, Tennessee on a site visit to the medical school, Otto Schmidt and I sat in a small cabin provided to visitors and debated the issue. We tried every combination of medical, clinical, and biology terminology with engineering and finally decided that the field was most suitably named "biomedical engineering" as representing most of the interests and the name stuck. The change was rapid. In 1967, Dr. James F Dickson and I wrote "Future Goals for Engineering in Biology and Medicine" because the name was not established and two years later, I wrote "Biomedical Engineering" and it was accepted as a normal title. Jim Dickson later added the current logo of the Society to help identify it.

There was a major problem in the field. There was no research support identified as biomedical engineering. Researchers could obtain training grants in biomedical engineering because a Biomedical Engineering Training Committee had been organized which clearly defined the field of training but no bioengineering research existed as such and grants were assigned by the Division of Research Grants to each institute on the basis of content. Many researchers engaged in applied engineering could not obtain grants because they did not fit neatly into the rubrics of NIH. Others who had grants from ONR (Office of Naval Research), NSF (National Science Foundation) or others, had difficulty convincing committees of NIH that their research applied to NIH interests.

At the same time NIH Institutes were doing biomedical engineering but under their own rubric rather than as biomedical engineering. The National Heart Institute was exploring the nuclear powered heart and DeBakey at Baylor under their sponsorship was exploring development of the artificial heart. Scotty Pratt at NIH was studying artificial intelligence and computer simulations. There were projects on artificial organs. The National Cancer Institute was working on radiation treatment patterns and intravenous drug administration devices. Considerable activity was going on elsewhere. A long range project in high pressure chromatography was in place in Johns Hopkins, and Norm Anderson at Oak Ridge was devising new analytical apparatus. Obviously, NASA was working on many projects in biomedical engineering, NSF supported a program in related fields, and DoD had projects on temperature loss and bionic controls. There was no coordination between institutes and with other government agencies and in fact a good deal of rivalry existed. It was almost impossible to find out what other units were doing or how projects overlapped.

Nevertheless, the field grew. Meetings were held at the 20 and 21st. annual meetings of the Conference on Engineering in Medicine and Biology in Boston and Houston in 1967-68. In the middle sixties, I called a meeting at the FASEB (Federation of American Societies for Experimental Biology) meetings in Atlantic City by passing announcements around asking those interested to attend. About fifty persons came to the meeting. Looking back, the meeting was totally male. Female engineers of any kind were not in great supply at the time. The purpose of the meeting was to form a society of biomedical engineers and it started with a great debate. Some felt that a society was not necessary; that the American Physiology Society and the engineering societies could perform all necessary functions, others felt that the society should be based in engineering rather than biology and some felt that a separate society was necessary. The latter view prevailed. The Biomedical Medical Engineering Society was formed and the organization incorporated under Illinois law in 1968. Otto Schmidt was elected as first president. We had no administrative mechanism so I ran the society out of my office at NIH for the first year and held the next election by mail-in ballot. It was embarrassing as I was nominated for office in the Society and had to count the ballots including my own votes. Later the office was moved to Northwestern University and John Jacobs undertook to manage it. The first dues were $10 per year.

I decided that identification of the field was critical. In order to rouse public interest I became editor of the *Advances in Biomedical Engineering* and later Don Gann and I wrote a two volume work, *Engineering Principles in Physiology* both for Academic Press.

With organization behind us, Dr. Donald Gann of Johns Hopkins University proposed to hold a meeting of the new society in Baltimore and at that meeting I brought in the editor of Academic Press with whom I had worked closely on several publications including *Advances in Biomedical Engineering*. He offered to support a journal and the *Annals of Biomedical Engineering* was born. The editor, John Lyman, from the University of Southern California began to solicit articles. It was tough going at first and it was several years before a backlog of articles began to occur. I served on the editorial board for some years.

By now, training in biomedical engineering had been supported by NIH for some time. A Symposium on training had been held in 1962 by the Institute of Radio Engineers (the forerunner of the IEEE, Institute of Electrical and Electronic

Engineering) but concentrated on engineering training without much regard for the biological sciences. A previous meeting of the Engineers Joint Council in 1961 had requested that I talk about engineering in biology as a discipline-I did so and had a long question period about the field with the general conclusion that training was needed. We requested and obtained money from Congress for the express purpose of training in biomedical engineering in 1963 and 1964, and in 1966 Dr. James Shannon, the long time NIH Director, indicated an interest in founding biomedical engineering research centers which lead to further development of the field. In rapid order some twenty or more training programs were established in universities, engineering schools, medical schools, and as multidiciplinary organizations.

It was time to develop a research base in Bioengineering. As a beginning the Committee of Consultants on Medical Research for the DHEW (now DHHR) had recommended attention be paid to the use of computers in medicine in 1960. The opportunity for development came when Dr. Stone and I were asked to move to the Division of Research Resources at NIH. In 1962 some 18 program in computer science were moved from DGMS to DRR along with John Jacob's Biomedical Engineering Resource at Northwestern. The division had previously occupied itself with large resources for basic sciences. There was an animal resource (primate centers), RMP (regional research centers for medical schools), and a Clinical Research Centers program. We now began to diversify into instrumentation and bioengineering. Norman Anderson at Oak Ridge developed major resources for analysis of clinical samples, later used by NASA on space flights. Marjorie and Evan Horning at Baylor received a large grant for development of mass spectrometers and many other awards were made in a similar vein. The budget was about $8 million per year for new projects in 1964 when both Stone and I moved back to General Medical Sciences (now an Institute).

Interest in biomedical engineering as an economic force began to develop. The Governor of South Dakota, Niles Boe and Otto Kerner, Governor of Illinois in 1966 and 1967 asked me to set up conferences for their offices to discuss the role of biomedical engineering in industrial development in those states. They were the first of many such visits.

World interest was also developing in biomedical engineering. In 1969 the Department of Commerce and the Department of State asked me to organize and chair a Technical Conference in Sydney, Australia to discuss innovations in the field. We pulled together several people from both Australia and the United States and persuaded the instrument manufacturers to provide exhibits. Large crowds attended. Later I went to Japan on a somewhat similar mission and had the opportunity to visit many Japanese instrument makers and see the new television developments being produced by Sony. Visits to France, Holland, Sweden and Great Britain followed.

Each country approached the problem in a different way and considered biomedical engineering in a different light. The Scots were interested in prosthetic devices, Wolfe in England was interested in the design of support systems for handicapped persons through design of appliances for the home not the person, Jacobsen and others in Stockholm were more theoretically oriented. Although the English still called the field Clinical Engineering or Medical Engineering it was part and parcel of the larger field.

In the early sixties we met with Orr Reynolds of NASA and discussed a possibility of design of a minicomputer.(Chapter 2). A group at Lincoln Labs (Jerry Cox, Wes Clark, Charles Molnar, and others) were interested, had developed a prototype and we made a joint contract with MIT. The work was not without trauma.

The dichotomy in biomedical engineering mentioned above is demonstrated by the organization of the field. The IEEE had maintained a professional group, a society, and related activities including a journal since 1952. It became focussed more directly on biomedical engineering in 1963 when the name was changed to the section on Biomedical Engineering. In 1964 it was merged with the old section on Medical Electronics (ETMB) and the name was changed to Group on Engineering in Biology and Medicine. I served as National Secretary in 1964. It became a Society of the IEEE in 1978.

Other developments were ongoing. The Alliance for Engineering in Biology and Medicine had also been in existence for a long time but again did not become clearly focussed until 1969. The AEMB has always been directed toward the clinical applications while the BMES has been oriented towards more theoretical fields. The AAMI (Association for the Advancement of Medical Instrumentation) has been interested in instrument development since its founding in 1965. The international biomedical engineering group is represented by an international society, the International Federation for Medical and Biological Engineering representing some 35 countries and organizations. It was formed in 1959. Several clinically related societies have also formed within recent years. These are members of the AEMB if they are technically related (Society for Artificial Organs) but others such as the Institute for Medical Records, Symposium on Computers Applied to Medical Care, etc. are new standalone groups. The truly basic scientific societies (The American Physiological Society), once almost enemies of biomedical engineering, now conduct symposia on use of computers in biological systems and systems engineering. I can recall being called to a meeting at Federation Headquarters in Washington to meet with the APS Council where I was advised that BMES should be melded with APS and go out of business. We refused. In later years Orr Reynolds became the Executive Director of APS and the mood changed to one of cooperation with engineering societies. BMES met with the FASEB for many years as a part but not a real partner in that large organization. Several meetings were held in the 60's and early 70's at which attempts were made to have the BMES recognized as a equal partner in the FASEB. FASEB always refused. It was a fortunate decision on the part of FASEB and BMES. The trend of biomedical engineering has been away from the basic sciences and instrumentation for biology to a stronger emphasis on systems and medical applications. My own career followed a similar path. My first articles were for the design of a new Warburg Respirator, then for a respiratory apparatus to partition human breathing, a study of system control of the endocrine system, and finally for new concepts of patient records and new computer memory design.

In 1970 I moved from NIH to the Office of Deputy Administrator of HSMHA (Health Services and Mental Health Administration) with responsibility for R and D and I turned my attention to engineering for the provision of health services.

During my tenure at HSMHA we worked on several projects that were allied to the broader aspects of biomedical engineering. We used a satellite, ATS-1 to provide medical information and treatment to Eskimos in remote sections of Alaska, we helped launch ATS-6 that provided mass telecommunications to India in many languages in programs related to health care and agricultural projects and worked on development of software and computers for patient records, diagnosis and treatment. In 1971 I made a presentation to the Domestic Council and obtained funds for the development of an Emergency Medical System (EMS) (Chapter 4).

A debate has always existed and is still going on regarding the scope of biomedical engineering. The biologists want to restrict it to basic science, the engineers to instrumentation, but new areas are emerging that are biomedical engineering but do not fit neatly into such categories. One is the application of engineering principles to the design of management systems for medical and hospital information. An example may suffice. I had worked with NASA on projects at NIH and now Sam Pool, Director of the Medical Applications Division approached me with the suggestion that we apply NASA engineering to medical problems using telecommunication as the main scientific input (Chapter 5). It was a new form of biomedical engineering.

We have recently been working with the Japanese on the development of a credit card which holds up to 4 million bytes of information and is formatted to provide a complete patient record which can be accessed by a personal computer wherever the patient may be. (Chapter 2)

Because most of the training and research programs in biomedical engineering are funded by the government, politics has always played an important role. It was difficult at first to convince Congress of the importance of the field. I remember appearing before an Appropriation committee with Daniel Flood as the Chairman with a set of amplifiers made for me by Wen Ko of Case Western Reserve that ranged from the bulky vacuum tubes to a chip to demonstrate progress (Chapter 13). A minority member on the same committee was Robert Michel (who is still in the Congress) who would lean over the long conference table and ask "well, what have you done for us THIS year." The demonstrations helped. We made enough of an impression that when Wilkes Barre was flooded in one of the major floods of the century. Flood requested that I come to help in the relief efforts. I knew nothing about flood relief but he was insistent so I spend several days in helping.

I served on the interagency advisory panel representing DHHS that selected the plans for operation of ATS-6. We met with Werner Van Braun, at Fairchild Corporation and examined the full scale ATS model. It was enormous. Great panes of solar cells projected from the large folding dish antenna. We were later able to go to Florida to see it launched from Cape Kennedy.

As biomedical engineering began to intrude in the field of delivery of health services, it came more and more to the notice of Congress because of financial and health reasons which were of interest to constituents. At one time the then Vice President informed me that NO money would be made available for EMS unless Baltimore was guaranteed a system. Senator Jennings from West Virginia demanded that a train be built and equipped to go through the mountains and deliver health care to miners as we had to the Indians. One well known senator threatened to block all centers for health research unless one was placed in HIS medical school. It was necessary to thread a close path between demands from all sides and the need for money for worthy projects. I hope we succeeded.

Recognition has always been a problem for the biomedical engineer. The field is not basic enough for the basic scientists and the engineers believe that the biomedical engineer is a chemical or electrical or mechanical engineer-not a biomedical engineer. This has been reflected in the honors system for science recognition. The first biomedical engineer to be elected to NAE was elected for his achievements in electrical engineering-not for Biomedicalengineering. It was not until 1987 that the NAE saw fit to create a panel to select members specifically for Biomedical engineering excellence. Before that time, election was by panels of specific types of engineers and it was often difficult to persuade a set-in-the-mold group that some areas of research called biomedical engineering were actually engineering. We were fortunate that the early NAE had foresighted members like Zworykin, Kinsell, Bolt, and others who recognized that Bioengineering would be a separate field and continued to press for support in the NAE.

I am proud that my election specified "Biomedical engineering theory and practice" as the reasons for election as the fourth engineer so selected. In the first 15 years of the society only 7 persons identified as biomedical engineers were elected; in the next five years 11 were selected and the last 10 years 18 were selected with 10 being selected in the last three years. There are now some 35 identified bioengineers in the Academy. Recognition is here and biomedical engineers are being recognized as a breed of engineers with the same status as chemical, or electrical or mechanical engineers. Great credit is due to Eugene Murphy, one of the first biomedical engineers elected to the NAE. He has always pushed for recognition of biomedical engineering and he and I have nominated many of the present biomedical engineer members in the Academy.

It was also difficult to obtain the status of Fellow in the national engineering societies for the same reasons. Only after the BMES section was established in the IEEE were biomedical engineers elected to Fellow grades and in my case it occurred 4 years after election to the NAE.

In 1990 Biomedical Engineering is strong and healthy. The membership of the Society is increasing and there are over a 1000 members, the International Federation has 7000 members and AEMB has 19 societies with many thousands of members. There are several journals of biomedical engineering often represented by the name of the specialty and a backlog of journal articles. The engineering directory of the AEMB lists approximately 200 schools which have a biomedical engineering training program(s). It has been difficult to identify the actual numbers of biomedical engineers which graduate but it is certainly in the hundreds.

We have come a great distance from a BMES with 76 members and the inability to fill a single journal issue with good research reports. The next 20 years are unpredictable but the fate of the discipline is secure.

Engineering and biology are changing rapidly. New demands on the health care system are no longer emphasizing costly new technology directed at more precise diagnosis or treatment. Today and for the foreseeable future the emphasis will be upon management of the system, reducing the cost of technology, and improving delivery of services. Innovations are desperately needed. Interactive computer systems, interconnections between parts of the system which are now isolated, care of patients by minimum monitoring in the home, devices which enable machine to replace expensive manpower rather than require more manpower as most do now, and the ability to transfer records and analysis from one location to another without the present problems all need immediate attention. We have been interested in the general problem from the project with NASA aimed at developing a total system to the optical card mentioned above which permits rapid transfer of complete patient records.

REFERENCES

Brown, J.H.U. and D. Gann: *Engineering Principles in Physiology*. Vol. 1 and 2, Academic Press, NY. 1972

Brown, J.H.U. and J.F. Dickson: *Advances in Biomedical Engineering*, Vols. 2,3 4,5,6,7,8. Academic Press, NY 1973-1979

CHAPTER 4

APPLIED BIOCHEMISTRY

Serendipity has played a large part in many of my researches. While at the University of North Carolina, I was working in endocrinology, particularly in adrenal function. As a part of the work, I researched the literature as every scientist must. In digging through old journals in the library I found two interesting articles. One in an agricultural journal reported that toxicity tests on animals revealed that there were changes in size of the adrenal gland when the animals were fed the insecticide, DDT (dichloro, diphenyl, trichloroethane) that was used in multi-ton lots world wide. The raw data was presented and no attempt was made to explain it. It was the obvious work of technicians who collected the data but made no attempt to understand it (Chapter 1). Secondly, John Gardner in the Department of Pathology at North Carolina also fed insecticides to dogs and found that the adrenals decreased in size.

This was an interesting observation but was not carried further although both the Department of Agriculture and Gardner had reported that the results were not consistent. I had a moment of serendipity. If there were variations in the effect of DDT on animals, could not the result be due to contaminants in the DDT that appeared in varying quantities in different batches and therefore caused the variation?

When I moved to Emory University in Atlanta the situation was favorable to look further into the matter. One of my graduate students was Cipriano Cueto, a chemist at the Center for Disease Control (CDC) in Atlanta. Cippy had obtained a master's degree in synthetic organic chemistry and wanted to work toward a Ph.D. We devised a series of experiments to test for the contaminants in DDT.

Dogs and rats were fed the DDT in varying doses and for varying times and the adrenals were examined in two ways. The steroid excretion, a measure of adrenal function was determined and the adrenals were examined histologically at autopsy. The chemical tests for adrenal steroids were not specific at the time but provided clues to adrenal behavior. The only really reliable tests were biological assays that were difficult and expensive.

In the meantime, Cippy fractionated large batches of DDD. He isolated components of the DDT and identified them by various means using spectroscopy, chemical degradation, and synthesis. As we had suspected, the DDT contained a variety of compounds in varying amounts. One of the most persistent byproducts was DDD (dichloroethane), a molecule containing two rather than the three chlorine atoms (trichloroethane) in DDT arranged in a particular form, the ortho-para positions. It was present in an amount of less than one percent by weight.

Further research was conducted by several graduate students who should obtain credit. S. Dutta was an Indian physician who came to us for research and an advanced degree while studying for medical examinations for admission to practice in the United States. He went on to become a high official in the FDA. Frank LaBella worked on the cytology of the adrenal gland and later on the separation of pituitary hormones. Frode Ulvedal, Atalante Petkas, Carlos Nunoz and Adolpho Portelo worked on the analytical problems. Frode went on to obtain the Ph.D. and become Assistant to Director in the EPA. Frank won the Abel Award for work started in the lab while Carlos and Adolpho came from South America to work in the lab as postdoctorals and returned to major positions.

We were now able to feed several different types of chemicals related to and isolated from DDT to animals. Two results were soon apparent. Rats were not a suitable subject. They appeared to not be effected by the DDT or its derivatives. We posited that this might be due to the fact that the rat produces corticosterone rather than hydrocortisone as the principle adrenal steroid.

On the other hand there were pronounced effects in the dog. The excretion of adrenal steroids by the adrenal cortex decreased, the glucose tolerance test which measures the ability of an animal to metabolize carbohydrates severely affected in adrenal cortical deficiency, was depressed. In addition, a critical observation was made. There appeared to be little or no upset in the ability of an animal to excrete salt and this again indicated a very specific site of attack for the DDT or its derivatives because the carbohydrate affecting steroids and the salt affecting steroids were presumed to be produced in two different regions of the adrenal cortex. If the experimental results could be duplicated we had found a method of treating one kind of adrenal disease without affecting another and of separating the exact location for the synthesis of steroids.

In a series of further tests, it was found that the effects were produced almost exclusively by the o,p-DDD. We prepared large amounts of the compound and performed a series of tests which confirmed our first results. On autopsy we found that the adrenal glands were severely damaged. The inner layers of the adrenal cortex which supposedly synthesized the steroids responsible for regulation of carbohydrate metabolism were destroyed and replaced by fibrous tissue. The outer layer of the cortex, the glomerulosa, which was presumed to be responsible for regulation of salt metabolism, appeared intact. This clearly demonstrated the independence of the layers for the first time.

The results were published almost at once in *Endocrinology*. At about the same time, a resident in endocrinology at the Emory University School of Medicine, Dr. Charles Harmon Brown, had gotten in touch with me. We had been analyzing blood and urine for pituitary and other hormones for the hospitals. He asked us to perform some analyses for him and we did so. In the course of his visits to the laboratory he became interested in the DDD experiments and analyzed the data carefully.

One day, Harmon appeared in the lab. He had a patient with an inoperable adrenal tumor which was producing large amounts of steroids. She was an enormously obese woman from the hills of North Carolina with high blood pressure and other problems which made surgery risky. We decided to try DDD. DDD was packed into capsules and a dose estimated based on the amount per kilogram which had been fed to dogs. The patient was fully informed and agreed to be a subject. She was fed DDD and the blood and urine were observed over a period of days. There was some nausea but nothing to prevent continuing the experiment. The adrenal steroids fell steadily. The patient's symptoms improved and she appeared on the way to recovery. She later died of complications from her heart disease. Autopsy revealed destruction of the inner layers of the adrenal cortex.

We wrote a short article on the clinical trial and submitted it to *Lancet* and it was accepted immediately. The publication in *Endocrinology* and *Lancet* caused an immediate response. We received letters from all over the world requesting samples of DDD.

I was a relatively naive researcher. A small grant from the Air Force (Chapter 8) and a small grant from NIH to investigate the DDD comprised my total research support. Grants were now received from the National Cancer Institute (NCI), the Damon Runyon Fund and others to continue the work. The Damon Runyon Fund people came with reporters, TV cameras, and all paraphernalia to record the presentation of a large check and for the first time my name appeared in the papers. Until this time my contacts with NIH were solely by letter. I had never been to Washington.

For this reason, I was greatly impressed to get a telephone call from Dr. Roy Hertz, Director of Research for the NCI (National Cancer Institute) asking about my research. He wanted to see the slides and the data and offered to come to Atlanta and in fact appeared the next day. Hertz was a small, meticulously dressed individual in a pin striped suit, highly shined shoes and metal rimmed spectacles. He was very quiet and looked more like a minister than a scientist. He looked at the slides and asked if I could produce a few pounds of DDD to be tested at NIH. Of course I complied.

Large samples of DDD were sent to NCI and they were parceled out to Grif Ross, then at NCI and later the Director of the Clinical Center and still later, Dean, University of Texas Medical School in Houston, and Dr. Bergenstahl in the Clinical Center who would conduct clinical tests. Over a period of months, our results were confirmed completely. Bergenstahl published a series of papers with many clinical cases of adrenal tumor successfully treated with DDD. I was nonplused when he failed to mention our work or to say from where his drug had come.

Over the next several years, we shipped many packages of DDD all over the world as we tried to comply with requests for the drug. For a time we operated a production facility and produced pounds of DDD from drums of DDT. Later two chemical companies began to produce DDD commercially under the trade names of *Orthane* and *Perthane* and papers began to appear confirming our results.

Two major incidents come to mind. One is that we were producing pounds of DDD with no real protection. The lab was full of DDT dust, the table tops were greasy from the compound. Considering the later determination of toxicity by the Agriculture Department and the banning of DDT as an insecticide in the United States, it is surprising that none of us suffered from toxic effects. Occasionally one or another would have a small bout of nausea which we attributed to the smell and not to toxicity. It was much later that the drug was banned.

The second event was the use of the compound as a drug. Nowadays, a new drug requires an intense testing program and approval schedule by the FDA (Food and Drug Administration) and clinical trials may require years and tens of millions of dollars to determine efficacy and toxicity. We notified the FDA of our intention to test the drug and then tested it. With the present restrictions, the drug would never have been tested in humans because we could not have afforded the clinical trials and a drug which is still being used for some forms of adrenal tumors might neve have been discovered.

By the same token, the use of animals was critical. There was no way in which the drug could have been isolated from a mixture of other molecules and then tested directly in human. As it was ineffective in rats and lower animals, the only choice was to use primates or dogs. Frankly, we would have preferred to use primates but they were too expensive, so dogs were selected as the second alternative and proved to be satisfactory. The antivivisectionists should take heed of a clear example of the need for experimental animals in medicine.

REFERENCES

Cueto, C. and J.H.U. Brown. *Biological Studies of an Adrenoorticolytic Drug*. Endocrinology 62: 334, 1958

Brown, J.H.U. *Ån Adrenolytic Drug* in Croxatto, A, *Pepidos Activos*, Press of Universidad Catolica, Santiago, Chile, 1959

CHAPTER 5

INDUSTRIAL RESEARCH

Nineteen hundred and forty eight was a bad year. We were broke, our baby had a brain defect and would shortly die, my wife had to spend time in Texas with the baby and her mother, and I was trying to finish a dissertation. Graduation was in June and in May I got a phone call from the Mellon Institute of Industrial Research in Pittsburgh. A component of the Institute, the Industrial Hygiene Foundation, was offering me a position as Director of Biological Research provided that I could come at once. The salary was more than could be made teaching and I needed the money. Without further ado the degree was taken *in absentia* and we moved to Pittsburgh.

The Institute was the largest industrial research institution in the country at that time. It occupied a large building with enormous pillars surrounding it on all sides and Mr. Mellon had insisted that each be cut from a single piece of stone. It looked more like a government office building than a research laboratory.

My immediate supervisor was Ted Hatch, a renowned investigator on environmental problems. He was small, thin, active with a large mustache. I rapidly found that there was no biological research and no laboratory. It was my job to start first one and then the other. While we planned, I watched environmental assessment going on in a very dirty Pittsburgh. The skies were often dark and the snow was discolored. I became interested in measurement of the extent of the smog and invented a gadget that pulled air through a filter and regularly changed filters on an hourly basis so that the amount of contaminants per unit of time could be measured. It was placed on a window outside the lab and measured the contamination for several years. There were days when the filter was weighed with a solid black deposit from the steel mills. Fortunately, those times are past, but so is the steel industry in Pittsburgh.

We rented an old stable, garage, and storage building from the Pittsburgh Women's Hospital and began to clear it out for use as a laboratory. The accumulations of more than one hundred years of disuse were discarded and the place was washed and painted. Air compressors, laboratory equipment, animal quarters and all of the other paraphernalia of a research establishment had to be installed, before we could get to work. It then became necessary to attract contracts.

Our first contract was to investigate phosphors, a chemical compound that produced the light and lined the interior of fluorescent light globes. Most companies had used beryllium as one of the components but it was now becoming apparent that beryllium was a serious cancer causing agent and broken fluorescent globes caused a serious health hazard because a few nanograms of beryllium in the air could effect the lungs. GE asked us to investigate new phosphors with calcium instead of beryllium as the active agent. They wished to be sure that the new agents were not as toxic before placing them on the market.

Over a period of time we set up exposure chambers, found a way to disperse the phosphors in the air as dust, discovered an electrostatic means of trapping the dust to avoid human exposure and to measure the amount of material to which rats were exposed. We exposed rats for varying period of time to different concentrations of phosphor dust and examined their lungs and other organs at autopsy. We employed Paul Gross, pathologist with the City of Pittsburgh to work with us to spot any toxic reactions.

We found several new bits of information. One was that certain sizes of dust appeared to deposit in the lungs more than other sizes, and secondly, that the dust appeared not only to be nontoxic, but was rapidly segregated by white blood cells and walled off from other tissues. The results were published in the *American Journal of Pathology* and attracted wide attention.

The item that caught our eye was the retention of dust in the lungs. Different sizes of dust particles were deposited in different regions of the respiratory tract. We attempted to resolve the question. The obvious answer was to attempt the experiment with exposures to a constant size of particle in each experiment. We obtained clay and various sizes of carbon particles that were non-toxic and exposed rats to the dust. The separation of particles into various sizes was a long and tedious process. The dust had to be wet, broken into fine articles and thoroughly mixed with water in tall containers. The dust settled slowly and at regular intervals that could be calculated, the top of the liquid was drawn off. The removed liquid was then allowed to settle and the process was repeated again and again. Gradually fractions could be obtained with a very high percentage of a given particle size. Sizes from 5 microns to a few tenths of a micron were obtained and checked for size and consistency by microscope and electron microscopy.

It was necessary to build a device to divide each breath of air exhaled by the animal into several parts to that we could obtain dust from the nose, trachea, and upper, middle and lower lung volumes. We designed and built a rotating device that took the air from each breath and diverted it sequentially into separate chambers where each breath could be analyzed for carbon dioxide that told us where the sample came from (the more carbon dioxide, the deeper the sample from the lung), and for dust that told us what fraction of the dust was retained and where. The machine was calibrated so that each subsequent breath diverted the same fraction of expired air into the same chamber. We obtained a regular curve of retention demonstrating that the large particles of dust were retained in the nose, dust about one micron in diameter was retained in the lung, and smaller sizes were simply breathed out again.

In the meantime, we had been in touch with the British Research Council that was interested in the retention of dust in coal miner's lung that caused "black lung". Application of our device to the problem might result in information that could aid in treatment or help to reduce exposure to the dust. We decided to test the theory in humans, obtained a grant for the project, borrowed an old iron lung used in polio treatment in order to regulate the rate and depth of breathing, contracted for a larger model of the respiratory partitioning device and tested the device on humans. We used clay dust in order to avoid possible toxic effects.

I had a number of graduate students by this time as a result of appointment to the biology faculty of the University of Pittsburgh. Ken Cook. Nancy Butler, Ray Scric, Marian Westrick, and Greg Ney served as operators and subjects after I had tested the apparatus on myself to be sure that there were no untoward effects. Over a period of time we obtained similar results to those in rats. The *American Public Health Association* heard of our results and asked us to report at the annual meeting as a keynote speech and then to publish the report in the Journal.

The *National Geographic Magazine* heard of the experiment and in an article on Pittsburgh made pictures of the lab. They insisted that Nancy, a young and good looking girl, be the victim in the iron lung even though she had not been an experimental subject. I was posed in white coat looking scientific (as the photographer asked). The paper on the retention of dust in the human lung was the first investigation of its type. Johns-Mansville Company asked us to use the device to test asbestos but we were unable to obtain satisfactory suspensions of asbestos particles because of their long needle shape and did not complete the experiment. Lacking the present knowledge of the toxicity of the chemical, we might have unfortunately been the first to find out that asbestos caused cancer of the lung.

The two big projects in the lab were the phosphor and the dust retention projects, but we were also kept busy on other projects. We worked on the toxicity of parathion, a dangerous insecticide, on the smog development in the Monagahela Valley around Pittsburgh that lead to the inversion in 1950, causing the death of many people in Donora. We analyzed samples from Donora, and visited the area. I can still recall the white houses with the paint turned a dirty yellow in color by the output from the steel mills. We worked with Dr. Kutcher, a famous opthalmologist on removal of harmful objects from the eye, using the rabbit as the experimental subject. We attempted to use radioisotopes made in the cyclotron by Dr. Allen on the University of Pittsburgh campus to trace phosphors in rat lung. Radioisotopes were not widely used at the time. Allen would irradiate a metal target to produce the desired isotope and give it to me in a bucket to return to the laboratory for purification. Later in my work at Oak Ridge (Chapter 12) I found that we had been taking unnecessary risks and in fact Dr. Allen and his coworkers had to have cataracts removed as a result of the exposure. The results were inconclusive partly because of our lack of knowledge of how to use this strange new material.

I was beginning to find the climate of industrial research confining. We were expected to pay for the laboratory from fees generated by jobs from our industrial clients. The clients were not interested in true research, they wanted the answer to a specific problem on a go-or-no-go basis. Often the problem was mundane, regarding toxicity of a material or some similar question.

Saved by the bell, I received a letter from the University of North Carolina offering me the position of Assistant Professor of Physiology in the Medical School. Although it paid less than Mellon, I accepted readily. There was the usual problem which occurs whenever a teacher moves to a new location. Greg Ney was finishing a Master's Degree. Ken Cook and Marian Westrick were finishing doctorates and it was necessary to find immediate sponsors although I continued to supervise and remained on the graduate committees until each finished the degree. The problems were solved and I moved.

REFERENCES

Gross, P., J.H.U. Brown, R. Scric, N. Butler, M Westrick. *Toxicity of Calcium Halophosphors* Archives of Internal Medicine 11: 465 (1955)

CHAPTER 6

EMERGENCY

Several hundred thousand people each year face serious emergencies and many hundreds of thousands die with heart attacks, accidents, and poisonings. The best remedy for emergencies is prevention and the second best is a good response team trained to respond properly to whatever emergency arises.

In 1974, Vernon Wilson, the Administrator of the Health services and Mental Health Administration (HSMHA) offered me a position as Associate Deputy Administrator. I was in a difficult position at the time (Chapter 13) and I accepted.

HSMHA with a budget of $2.5 billion and 25,000 employees was a hodgepodge of agencies ranging from HMOs(Health Maintenance Organizations), the Indian Health Service, Health Financing, to Regional Medical Programs and the National Institute for Mental Health. I worked closely with Robert Laur, the Deputy Administrator. After assuming our duties the first major task came directly from the President, Richard Nixon. He asked for initiatives for a new program to rouse interest in health and attract support for the administration. I was given the task of formulating the program. Over a period of a few weeks we came up with several ideas. One was the centerpiece-the Emergency Medical Systems Program (EMS).

The outline that proposed a nationwide plan of emergency vehicles, hospitals, communications and management, was sent to the President. I received a call a few days later to appear in the old gray edifice next to the White House, the Old Executive Office Building, to present and defend my plan to the President's Domestic Council.

It was with some trepidation that I appeared at the specified time. The Executive Office Building was a mammoth structure erected to house the War Department during the Civil War. It had high ceilings, brass door knobs, and marble floors. Each room was approached by a huge dark door with a skylight over the transom. The atmosphere was intimidating. I was admitted to a large room with an enormous table. Seated around it were the members of the Domestic Council. The room was hot, smoky, and crowded. Squeezed into a chair in the back of the room I was too flustered to remember faces and the entire event is a blur. On the chairman's request I made a short presentation, answered a couple of questions largely about estimated costs, and was told to leave. Some days later I was advised that the Council had accepted the plan, and would include funds in the budget, but that the program would be restricted to no more than six sites as test beds with the possibility of further expansion later.

We immediately created the Emergency Medical Services Program and ran it from my office for a few months while we attempted to get the idea off the ground and recruit personnel. Tam Yolles, a physician and wife of wife of Stan Yolles, the former Director of the National Institute of Mental Health came from Long Island University Medical College of the State of New York to serve part time. John Reardon was attracted from a think tank in the Washington area, and David Boyd, head of the Emergency Services for Chicago, came to head the division when it became formally established.

An announcement was made that proposals would be accepted for contracts with cities or states or other political entities to establish emergency services. We received dozens of applications. A committee of physicians was set up to review and make recommendations on the finalists. Over the next few months we crossed the country several times reviewing the most likely applicants. We traveled to the wilds of New Mexico, the forests of New Hampshire, and the Northwest as well as to the majority of the large cities in the country.

Criteria were difficult to establish. We had conceived of a total system that would encompass ambulances, monitoring from ambulances, training of paramedics to provide medical services, dispatch services with a single telephone number (911 was adopted), emergency suites in hospitals available on 24 hour service with trained emergency physicians, referral services for drugs, alcoholism, and poisonings, psychiatric counseling and a system director. No system existed.

The diversity of the applications that we received was great. There were applications to set up ambulance systems, Arizona wanted to patrol the highways to handle traffic accidents, a city was interested in switching the ambulances from the control of the police to the fire department. Dozen of other variations were seen, none representing a good system.

On review of the applications we found no total system in existence. St. Louis had as many as 20 different numbers to call for emergency services. Florida had an excellent ambulance service but almost no hospital care. Some cities had ambulance service consisting of the hearses from funeral homes. Some funeral homes had contracts for service in specific areas and if an accident occurred across the street from their district, they refused to pick up victims. If the definition of an ideal system as outlined above was the criteria for a Grade 1 system, no such system existed in the United States. We had several million dollars available to establish systems and we wanted to create emergency services as near to a Grade 1 as possible. The objectives were not met during the course of the initial experiment.

For several months we visited parts of the United States from New Hampshire to California and from Illinois to Florida, reviewing applications, studying the systems in existence and picking the best parts of many to adapt to a total system. Although we had a very competent group of surgeons on the review team, we had no one who was specifically an emergency physician. I recall one physician, Truman Blocker, brother of Dan Blocker of Ponderosa television fame, Dean at the University of Texas Medical School in Galveston. Truman was huge like Dan. He towered over everyone. The taxi sagged when he entered. But he was a surgeon of great renown, a great administrator and an affable travelling companion. Another huge man was John Najarian from the University of Minnesota who later went on to organ transplant fame. The physician who had treated President John Kennedy at Parkland Hospital in Dallas after the shooting was also on the panel. The main problem for the committee was to select from the applications those that were most likely to be viable for improvement and upgrading. The choice was deliberately made to cover a variety of different systems to test the development of emergency care in various situations. We tried to locate suitable county, state, and local systems, systems which some elements in place and rural and urban area system proposals. For example, we funded a system in Ohio where several small towns were linked together with communications and ambulances to provide rural emergency care. The ambulances were based in rural areas with an extensive communication system so they could be in contact with the surrounding region. The State of Illinois proposed a system in which the hospitals of the state would be graded in ability to respond to emergencies and ambulances and helicopters would be positioned to transport victims from an accident to the most suitable locations where certain hospitals were designated as head trauma centers, others as heart centers or orthopedic centers. The entire system was linked with extensive communications. The City of San Diego was chosen because it had already been developing a system to establish medical care for the farming region of the central valleys of California as well as a local network with a very good poison treatment center.

The list of acceptable sites and the costs for the first year was sent to the White House. A call came from William Magruder, Special Counsel to the President, within a day or so asking us to come down and talk to him. We sat in his office while he hedged around the subject. The reason was soon obvious. Unless Baltimore was selected as one of the cities, Spiro Agnew, former Governor of Maryland, and Mayor of Baltimore, and now Vice President of the United States, would see that the program died.

Baltimore had one of the nation's best emergency transport systems. Adams Cowley, of the University of Maryland, had developed a system of using the police helicopters to pick up highway injuries which he protected jealously. He operated the system as he saw fit without regard to the city or other support. Baltimore had been turned down for a grant because we could not see funding him and allowing a one man operation. We now went to Baltimore, talked with the representatives of Johns Hopkins University and the City and finally developed a program for Baltimore as a whole headed by Donald Gann of Johns Hopkins University, Department of Surgery, which was acceptable to the advisory committee. The Emergency Medical Services was born.

Once the program began to develop, pressure arose in Congress from the disappointed applicants to create a program of Emergency Medical Services and a separate division was organized within HSMHA. The program was designed to upgrade emergency service with whatever was required. This might be communications, better training of paramedics, training of emergency physicians, improving hospital facilities, or other measures. Over the next few years several hundred millions of dollars were spent. As a result the EMS of the country is now in much better condition. Most large cities have adequate ambulance service staffed by trained paramedics, many have helicopter services, and emergency suites are headed by trained professionals as training was developed for emergency medical personnel. Still the number of Grade 1 facilities is very limited. For example, in the city of Houston with 4.5 million inhabitants, there is only one facility classed as Grade 1.

Because of previous connections with NIH, we were able to encourage development of emergency physician training programs in the Surgery Training Section and to develop programs of training paramedics. The ambulance attendants had previously been attendants with little or no medical experience. In cases where hearses were used as ambulance, the attendant was often the driver. The program which was envisioned and which gradually developed was one of several hundred hours of training in medical procedures creating the paramedics which are now an established feature of most ambulance services. Now the paramedic is often especially trained in treating heart attacks or trauma. The ambulances were upgraded by creating designs for efficient operation and most of the high, box shaped ambulance now seen on city streets originated here. Residencies were established in medical schools in emergency medicine and the Emergency Physicians Medical Association was organized. Journals dealing with emergency medicine were established. In 1978 I received an award from the journal *Emergency* as EMS Leader.

There were many problems. Few states required by law any Grade 1 emergency suites and administrative problems plagued the system. California, for example, passed a law which said that an ambulance would take a victim to the nearest hospital regardless of the qualifications of the hospital to treat the emergency. As the system grew, every hospital wanted to have an emergency room and competed for business because most emergencies resulted in admission to the hospital and thus increased revenue. Hospitals advertised their emergency capability, often overstating their ability to treat emergencies. Jacksonville, for example, had an ambulance system managed by a former Naval officer with the best response time in the country. The number of fatalities which occurred during transport was very small. However, when we examined the system we found that fatalities had dropped in the ambulances but had increased in the hospital indicating that the hospital did not have satisfactory treatment and that there was no real emergency system.

The implementation of a 911 emergency number was very difficult. The 911 number serviced all elements of the response system, fire, police, and emergency. The political ramifications of coordination between the departments of a large city to select frequencies for radio transmission, to determine who would provide dispatch services and manage the system created major problems which required years of negotiations in some cities. The highly trained paramedics were also placed at a disadvantage. Although they were detailed solely to ambulances, promotion could be obtained only by returning to fire fighting and fulfilling the requirements of a firefighter before promotion. Many of the problems are still unsolved. Some cities do not yet have 911 or have recently installed it, promotion of paramedics is still under the control of the fire chief, and many cities do not have a Grade 1 facility.

Although the EMS was a success despite the caveats mentioned above, Congress gradually ceased funding as other priorities became more prominent and the division went out of business in 1987. It was left to the cities and states to pick up the bill. Perversely, with the current decrease in Federal funding many cities are finding that it is necessary to decrease emergency services and the grade of service provided may decrease. In Houston in 1992, the Grade 1 facility at Ben Taub Hospital is now being bypassed to take victims to less well equipped hospitals because of overcrowding.

Another project closely related to the EMS system was a disaster, although the idea was a sound one. One of our proposals to the President was to establish a central point for Federal funding. All funds for human services would be channeled to this point and the distribution would be decided by the local community. This would permit the use of funds in the most efficient manner, decrease the overhead costs and personnel needed for administration and improve health care in a community. The idea was called the Experimental Health Services Delivery System (EHSDS). The federal funding of social and health services was to be funneled into the EHSDS and then to projects on the basis of merit. The parochial attitude of both local and federal authorities doomed the project. No one wanted to surrender the prerogative of funding a project and obtaining personal credit for it. The independent fund raising organizations such as the American Heart Association, the American Cancer Association, and others, could see their funds submerged into a pool of unidentifiable projects and thus decrease the chance to raise funds. The nearest approach to a viable system was in Wilkes Barre, Pennsylvania.

The Congressional representative from Wilkes Barre was Daniel Flood. Dan Flood had been an actor and clearly demonstrated it. He appeared in hearings clad in an ascot tie and loudly checked sport coat. He wore a carefully waxed handlebar mustache which he stroked continually. A Chairman of the Appropriations Committee for Health in the House of Representative, he wielded great power and did not hesitate to use it. He was fond of quoting Shakespeare to back up his points.

Flood was interested in the EHSDS idea and tried to implement it in Wilkes Barre. It was a partial success helped by a natural disaster. Wilkes Barre had a large flood in 1973 and the EHSDS was used to coordinate rescue and provide services. The river broke a levee in Wilkes Barre and enormous amount of water spilled into the town. Houses had six inches of mud on the floors, streets were buried to the curbs, and everything was destroyed. There was a major problem of providing food, establishing communications, and preventing illness. I was sent to Wilkes Barre at the direct request of Dan Flood and I slept next to him in a hanger at the airfield (although his home was nearby, he stayed with the rescue effort) for several days while we tried to coordinate emergency procedures. It was my first real emergency.

We also tried to implement a system in Tucson using an instrument already in place, the Health Council which had been established by law to administer or recommend on government programs. It was not a success because of the parochial attitude of many of the community fathers.

The Health Council was a health service agency at the local level established with a composition of equal parts health care providers and civilians with an interest in the health care system. It was supposed to review all major expenditure of funds and decide what was good for the community. In turn, if a project was turned down, no Federal funds could go into it. The meetings were a farce. I have walked in the door to hear two individuals talking a corner, "You support my CAT scanner, and I'll vote for your new wing." In addition, the members of the panel were often not well versed in medical costs and methods. I watched a physician appeal for funds to transmit EEG (electroencephalogram, brain waves) to Chicago to be read by computer and returned to him. He failed to mention that such a procedure requires high volume of patients and that he could not obtain such volume in a community of 3500 souls. But the project was approved because it enhanced communication. Many such examples could be cited.

The experience with the emergency medical services system clearly demonstrated that the system was more political than medical. The sociological aspects of coordination between fire and police, the resistance of hospitals to change, and the desire of politicians to "get on the band wagon" created resistance to the creation of a good working system. I remember Rizzo, the mayor of Philadelphia, threatening to go to Congress or the President if he did not get a system. One the other hand, there were some satisfactions. Long after I left Washington, I received a telegram from the White House, signed by Gerald Ford, asking me to attend a meeting on EMS. The university in which I was located at the time was greatly impressed, but the meeting was largely rhetoric. When Dan Flood requested my presence at the flood in Wilkes Barre, I was flown on the White House Flight and had a chance to collect napkins, matches and other small items with the Presidential seal. There is no greater thrill than arriving at Andrews Air Force Base, being ushered into the VIP lounge, escorted to the plane with United States of America painted large upon it, seated by a white coated steward, offered drinks and food, and sent by car and helicopter to one's destination after arrival. It is one perquisite missed in an academic atmosphere.

REFERENCES

Brown, J.H.U., *The Crisis Emergency System.* Emergency Medical Services, 6: 58 (1977).

Brown, J.H.U. *Emergency Care*, Texas Medicine, 75: 63, (1978)

CHAPTER 7

APPLICATIONS OF SPACE SCIENCE

NASA was placed under considerable pressure in the 1970's. Men had been to the moon but Congress was now asking for tangible earthly results. The Congressional committees wanted to see the technology invented for space applied on earth. NASA had mounted a program with the universities to identify technology that could be used in civilian pursuits but it had not been uniformly successful. There seemed to be no way in which technology designed for the space program could be applied on earth except for rare instances. NASA had gone so far as to propose the moon buggy used for travel on the lunar surface be promoted for use of the handi-capped. It was an expensive proposal for a minor improvement in status and this was typical of most proposals.

As a result in 1973, NASA decided to exploit the use and transmission of data from remote locations to provide for social purposes. It was decided that the most likely exploitation of value to both NASA and science would be in the use of advanced communications in health care. This would help the emergency systems (Chapter 6) by providing communications over a wide region and would also enable NASA to learn something about the care of patients with remote services that could be applied in the case of an astronaut's accident or illness in space.

At the time NIH, the National Library of Medicine and the Lister Hill Center for Medical Communications were using a NASA satellite poised over Alaska (ATS-1) which was providing communication between remote Eskimo villages of Northern Alaska and the mainland for medical advice and general communication. The antenna were placed in school yards and the transmission by satellite provided health information, a means of calling for transportation in emergencies, and for communications in general during the long arctic winter. The problem can be illustrated by a single example. Fort Wainwright was an Eskimo village on the Arctic Ocean, far north of the Arctic Circle. It was a hunter-gatherer society in a small community with whale bones piled along the beach. The only access was to fly in a light plane from Point Barrow for hours across the tundra to a gravel landing strip. There were no roads. The inhabitants were much more familiar with air travel than with any other form of mass transportation. Most of the inhabitants had never seen a car or a train. All communications were cut during the winter except for radio, but the satellite placed the remote location in contact with Anchorage with good health facilities.

The chief of the small settlement was Homer Bodfish. Homer had a lung removed in Anchorage for tuberculosis. When asked how he was, he responded,"I used to be able to chase down a wounded caribou but now I get tired after running for eight to nine miles." His reply illustrated perfectly the society and the hardship.

The radio operators in the typical Eskimo village, the teacher or someone willing to take on the tasks were given a few hours training in general first aid and how to identify more serious conditions. Their main function was to provide a contact to call for transportation or advice. The satellite provided voice communication only but it proved invaluable in providing help and calling for transportation in serious emergencies.

Now NASA proposed to launch another satellite (ATS-6) which would have television as well as voice channels. I served on the interagency advisory panel representing DHHS which selected the plans for operation of ATS-6. We met with Werner Van Braun, at Fairchild Corporation and examined the full scale ATS model. It was enormous. Great panes of solar cells projected from the large folding dish antenna. We were later able to go to Florida to see it launched from Cape Kennedy. It had many channels, and was much more powerful than ATS-1. The satellite was first situated to produce a foot print or coverage to provide medical information between Alaska and WAMI (a consortium of the western states of Washington, Alaska, Montana, Idaho) to train medical students in Alaska and the other states which had no medical school by connecting them to the medical school in Washington state. ATS-6 was then to be positioned over Africa (1975) to provide television transmissions to 5,000 villages in India for agricultural advice, birth control lectures, nutrition demonstrations and similar uses. There was a major logistical problem in that many languages had to be programmed and the information had to be carefully screened for ethnic and religious attitudes. The experiment was not an unqualified success. Evaluation revealed that the programs were watched mostly by the women who had little to say in agriculture or birth control, and that it was more a source of entertainment than instruction. Satellites are a major means of communication but they suffer from several faults. Both launch and maintenance are expensive. ATS-1 was shut down because NASA could not afford the one million dollars per year to keep it operating for health care only and DHHS refused to fund the project. Operations required stations on the ground, operators, calculations to adjust orbits, and many other expensive details. In addition, the cost of launch was in the tens of millions of dollars and if the satellite was to be used for medical experiments, the chance of recovering launch costs was not great. Commercial satellites could be used to transmit medical data if funds existed to pay the high costs and if the footprint of the satellite reached the desired locations.

In 1974, Sam Pool and Norman Bęlasco from NASA came to see me. They proposed to establish a complete health delivery system which would depend upon communications and they had already thought of the possibility of a remote location on earth in which to test it. NASA had developed IMBLMS (Integrated Medical and Behavioral Laboratory Measurement System) for long duration experiments in orbit and they wished to test it in a real environment We had a solution. As the Associate Deputy Administrator of HSMHA, one of my oversight responsibilities was the Indian Health Service. NASA proposed to enlist the tribes in Arizona which were far from medical care, isolated in mountainous terrain in small villages with poor transportation and provide them with health care. It was a good idea and we proceeded to talk to the director of the Indian Health Service, Emory Johnson and then to the central research arm of the health service in Tucson. The research director, Erwin (Stu) Rabeau was enthusiastic. We were in business.

STARPAHC (Space Technology Applied to Papago Advanced Health Care) resulted. A site selection board was appointed in which I represented DHHS. Four sites were considered: Alaska, the Navajo Indian Reservation, the Pueblo Reservation and the Papago Reservation. The sites were selected on the basis of remoteness, need for health care, possibility of providing remote health care, size of the community (no more than 10,000 individuals) and cooperation of the community. The Papago site was selected as fulfilling most of the criteria. In addition, it was close to the Indian Health Service research center in Tucson.

NASA proposed to build a van equipped with medical test equipment (ECG, X-ray, laboratory, and other equipment) using equipment which had been designed for compactness in order to be suitable for space flight. The van would travel from village to remote village, providing health care under the aegis of a paramedic who was always connected to physicians in the Tucson office by radio and television links directly from the van. Telecommunication was considered to be a vital link for transmitting x-rays, pictures of the patients and laboratory data from the remote van to Tucson and on to Phoenix where a general hospital was located. This then provided NASA with a test of medical services rendered in a remote location (space ?) by a paramedic with advice and followup provided by telecommunications.

It was necessary to start from scratch. Equipment, control stations, dishes for relay services, all had to be provided and personnel trained. The paramedic would be isolated from physicians and supplied with a variety of equipment. It was necessary that he be trained in clinical anatomy and physiology, basic dentistry, treatment of infections, basic laboratory skills, preparation of the medical history, diet and nutrition, elementary obstretrics and gynecology, elementary pediatrics, some pharmacology, radiology, social services and ophthalmology. As the technician was isolated and not under direct supervision, the personality of the individual was critical. One technician was dismissed for overzealous treatment and the "god" complex.

The tribal councils had to be brought into the scheme. Every activity on the reservation had to be approved by the council. This was a complex situation. Many mountains in Arizona are sacred to the Indians and relay stations could not be built on them. The route of the van had to be determined and approved. The Council had to convince the users that the system was safe and good for health.

Communications links had to be established. Mountains in Arizona interfered with high frequency radio and television and relay stations had to be constructed on mountain tops. The van had to be equipped and tested. The testing of equipment had to be from the ground up because there was no information on how x-rays could be transmitted and read accurately in a remote location. Experiments were conducted on the number of lines of raster, the resolution and the readability of the screens before deciding on a system. In some cases it was necessary to provide generators and concrete pads for parking in areas where no services existed. The van provided services as needed. Previously (at NIH) I had worked closely with

Norman Anderson at Oak Ridge to develop a small portable analyzer for body fluids and this was now incorporated into the van. The total system was installed and for some months it traveled from location to location on a regular schedule. It was evaluated in 1975-76 to determine the effectiveness. Interesting results developed.

We found that much of the elaborate equipment was not needed. The x-ray was rarely used; a stethoscope with electronic sound transmission had been installed but was never used. Video was not necessary except under rare circumstances. Simple medical care could be easily provided by the system. The conclusion was that the medical needs of the population should be considered and the minimum technology applied to satisfy the needs.

The system was a great success from another viewpoint. Many more patients were seen, patients received better care, the need to transport patients long distances for care was avoided. The patients were highly pleased and the tribal council was interested in maintaining the system. At the end of the evaluation period, NASA pulled out of the project as they had tested the feasibility of the idea but under their mandate they could not continue to support medical care. I wrote many letters to the Secretary asking for DHHS (Department of Health and Human Services) to support the system but the petitions were refused. In the end the project was disbanded. DHHS had contributed the time and effort of Indian Health Service personnel, the space and locations and had introduced us to the tribal councils which had to approve the project but they contributed no money. NASA contributed the equipment, operating costs, and expended large sums of money.

The project came to an end. The Smithsonian Institution requested and obtained the van for its collection. Several countries came to see the installation to test the feasibility for adoption in their own countries. Saudi Arabia was one. Many of the ideas were adopted by the tribal councils and implemented without the van. Better communications were established, more clinics were built and more paramedics were trained. The state of Alaska launched its own satellite. Overall, the system was a success.

REFERENCES

Brown, J.H.U., *Telecommunication for Health Care*, CRC Press, Boca Raton, Fl., 1982.

CHAPTER 8

RESEARCHING TOGETHER

Scientists are of two types: those whom by choice work alone and those who work in groups. The two types have different personalities and different approaches to science. The loner has a focussed approach to a problem. It is attacked piece by piece, often slowly and deliberately, trying to seek out the clues that will enable one to reach a conclusion, perhaps in secret, trying to conceal data from others. The group worker works piecemeal, parceling out aspects of a problem to close knit members of a group, each with separate expertise, in the hope of achieving a major breakthrough.

There is another approach. This is to link together expertise from widely disparate groups to attack a practical problem in the hope of reaching a useful end. With a pragmatic approach to science, I have been interested in this method. My major interest has always been the relationship between academia, industry and government. Cooperation between the trio, except for the research grants issued to universities by the government and the industrial-military complex, has been notably lacking. In order to succeed in the present competitive world a closer working relationship is necessary. Japan realized this long ago and the MITI (Ministry for International Trade and Industry) has worked to establish close links between government and industry and the university. We in the United states, have traditionally avoided such contacts to our disadvantage. The typical university professor in sciences is an advocate of "basic" sciences for basic science's sake regardless of the possible benefit or lack of it to society. The theory of organization of consortia has always interested me and appealed to my sense of the need for greater cooperation between all elements of society.

One way to encourage industrial-academic cooperation is the research consortium. The research consortium is a group of private, for profit, governmental or academic institutions that are organized in any combination to achieve a common goal. Most modern consortia have a single goal in mind-the transfer of technology from basic research to applied purposes. Such a transfer suggests that industry must be involved in the process in order to achieve a practical outcome, and, as a result, most consortia now involve business interests.

Thus the consortium is a means of achieving a direction of research that can diect basic ideas into practical application. Industry and academia must serve as equal partners in order for the operation to succeed. The consortium provides the vital link of communication between the laboratory and production. It is essential to develop the link in our present situation where the United States lags behind other industrialized countries in the ability to exploit innovation. The United States leads the world in innovation and scientific research, but trails many other countries in the Orient and Europe in the development of research into useful products. The consortium can provide the link.

It has not been so many years ago that the research consortium was confined mainly to academic institutions. In the usual arrangement a group of institutions established common library or support facilities. The HARLIC (Houston Area Research Library Consortium), which is composed of a large group of research libraries in the Houston Area is an example. Many institutions established joint efforts involving purchasing departments or laundry services. Departments would often collaborate with other departments. But relationships between industry and the university were kept at arms length. In fact, faculty members resented being labeled as "practical" and as doing "industrial" research. The government was equally suspect, and Department of Defense contracts were refused by many institutions. In many places only engineering schools maintained contact between the two alien camps.

Consortia are now being redefined and the barriers between industry and the university are coming down. Faculty now eagerly seek industrial contacts, patents are becoming a significant part of university income and contracts with industry are common place. Organization of consortia between government, the universities, and industry is accepted as an everyday event. But these relationships create strains on the part of both industry and the university.

Consortia must be organized for a purpose. That purpose is innovation and the transfer of technology. It should be emphasized that collaboration between individuals and/or departments and industry does not constitute a consortium and neither does the accumulation of industrial research in a research park. Although some consortia have general aims such as interlibrary cooperation, this does not constitute a research base or a way to transfer technology.

The brief definition of a consortium leads one immediately to the problem of organization. A consortium can be organized only if certain perquisites are present. The academic faculty must be engaged in research which shows some promise of application in technology. Industry(s) must be aware of the research and see practical implications for the future. A leader must be available who can see the connections between the two groups, establish communications between them, and convince each of the advantage of working together. And finally, both industry and the university must see benefits on a long or short term basis.

There are three major elements in any consortium: the scientific base, the funding and the administration. Like a three legged stool, each is equally important although the science must precede the other two. Without a strong base in science which can attract investors and without a clear direction, no consortium can succeed. As a result the track record and reputation of the science base is critical. All too often, the personnel which would be ideal are already established and have their own grants and support. Recruitment is hard. Choice of a director becomes a major task of any consortium. A Nobel Prize winner is more likely to be listened to when a new invention is discussed but is not easy to move. Management is important. Every element concerned must have confidence in the management and its ability to put together facilities, ideas, personnel and money to create a viable product.

There are four qualities to be determined in organizing a consortia for technological transfer. They are the quality of the management team, availability of funds, quality of the technology mentioned above and, finally, the image of success displayed by the consortium. Obviously the most important element in the beginning is the technology.

But consortia have problems related to both university and industry attitudes. The faculty are often not interested in participating in consortia although their expertise may be essential. Multiple loyalties create problems. Many professors and some industries tend to say "I can do it myself. Why should I share with others?" Many academic institutions fear the loss of autonomy especially if they are engaged in a consortium with multimillion or billion dollar industries. Faculty members are unwilling to enter into a contract until funding is assured while industry is unwilling to fund research until the services of experts are available; this chicken and the egg problem may be insoluble in some cases. Many industries and some universities are wary of government membership in a consortium. At one time, the major problem revolved around the open patent policy of the Federal government, but much of this problem has been resolved by new regulations. Faculty may not want to work on contracts which involve defense or toxic materials. Many institutions do not realize that the institution owes as much to the consortium as the consortium does to the institution.`

Strains within the university are often evident at the administrative and faculty level. The administration is usually willing to accept funds for work to be conducted by a faculty member but there are attendant problems. Some faculty members, in the fine arts for example, resent seeing an engineering professor with a fine income and other amenities as a result of a lucrative patent. The university may try to be even-handed, but it is difficult when one faculty member is bringing in large royalties to the university and another is using university funds to support small projects.

Universities have difficulties in handling personnel engaged in patentable research. They are now required to be much more astute in negotiating rights to inventions, in making clear the responsibility and rewards to the faculty member who develops a new process, and in spelling out terms of agreements. Suits by graduate students or fellows who may have participated in the research but gained inadequate financial rewards are ever more common.

Industry on the other hand has problems in dealing with the university. The industry wishing to exploit an idea or invention made in the university laboratory has a problem in securing rights when the university is engaged in a consortium which may involve several industries, each of which may lay claim to the same information. Industry may be asked to contribute to the support of a consortium and yet obtain no tangible results for some years as a result of the collaboration When the bottom line is critical to the shareholder, the industry may press for more rapid solutions to its problems.

The difficulties of the consortium relationship has resulted in some universities refusing to accept large industrial connections and in some industries refusing to join a consortium for fear of losing proprietary interest. Nevertheless, the collaborative effort has grown and dozens of consortium are now organized for purposes ranging from information exchange to development of inventions.

Despite the problems there are many advantages to cooperation. The consortium can draw upon university basic sciences for ideas for future development, use research dollars more efficiently by pooling information, avoid duplication of effort and undertake large scale expensive projects which could not be easily handled alone. In addition, the consortium allows a company to obtain royalty free patents from the consortium, learn about new disciplines which may impinge upon profits in the future and provide a source of employees from graduate students working on the projects.

The university also gains benefits from the consortium in exploiting a faculty member's idea for monetary return, and obtaining funds for support and development of the university. It should be emphasized again that collaboration between individuals and/or departments and industry does not constitute a consortium and neither does the accumulation of industrial research in a research park. The consortium also offers the chance to obtain expensive equipment and space and to become closer to the community and its aims.

A not inconsiderable return is the ability to attract and pay graduate students, attract new and expensive personnel, increase the number and quality of publications, and thereby enhance the university reputation. The advantages are counterbalanced by disadvantages. As an example, the government has become interested in the consortium as a trust entity and in fact has initiated anti-trust action against SRC for its organization of the semiconductor manufacturers into a single consortium to develop new devices. While it is true that a consortium, especially if industry is heavily involved, may combine most of the total resources in a given field into a single monolithic structure, this is not usually the case and the Congress has passed the National Cooperative Research Act to encourage joint efforts between industry and academia and between industries. However, the Justice Department and the Cogress have not yet determined where a trust action may lie.

Disadvantages also lie closer to home. The university may find that a consortium will divert loyalty from a school or department to the consortium, increase the demand on faculty for teaching and research, result in research oriented faculty doing less teaching and forcing graduate students into industrial research. It may also create conflicts of interests between industrial and university goals. A very real problem lies in promotion and tenure of staff who work in the consortium and may not teach a full time load.

The industry may find that disadvantages occur where there are no exclusive rights to development of a device, where support may be provided for periods of time without visible immediate returns, and where individual components of the consortium may be in competition with each other to develop a product from a nonexclusive license and trade secrets may be hard to maintain.

On the other hand the industrial member of a consortium may find that it has access to information and research which would not otherwise be available, that it can establish a research base at lower costs than total funding, that it may realize a profit from patents developed in the consortium, and that it may find basic ideas which can be applied in other areas.

The way in which consortia are organized are as varied as the institutions in which they are established. Some institutions such as MIT have elected to spin off small companies which exploit an idea and return a profit to the university which amounts to many millions of dollars per year. Other consortia have established nonprofit institutes in which industry, government and academia collaborate while others have maintained a direct university connection. The use of venture capital to spin off companies for development is rising steadily, while industry has also established separate facilities for consortia in some instances. Nevertheless, the problems are far outweighed by the advantage of consortium formation. Despite proprietary concerns companies have demonstrated great interest in joining groups interested in a single problem.

A consortium is organized for many purposes. Although there may be a main objective, the well run group may perform many functions. As a general rule the consortium will provide advice and information such as library consortia. It may coordinate activities of a diverse group toward a common objective. MITI in Japan has handled this function admirably. A group on solar energy development, semiconductor development, etc. are other examples.

The consortium may exist to provide information. The Oak Ridge Institute for Nuclear Studies is a consortium of some 50 universities which serves largely as an information exchange on nuclear problems and possibilities.

The consortium can foster development of technical expertise development. This is probably the major thrust of consortia. An obvious role is the provision of funds. The major thrust of the supercollider consortia appealing to Congress for funds is the $6 billion which will be provided to the economy of the state in which it is to be located. A consortium may pull together disparate research efforts into a common focus. Regardless of the organization and the funding, a modern consortium should have a major goal in mind--the transfer of technology from a basic idea or principle into practical application. Technology transfer, to be discussed below, is critical to success with industry because unless there is a pay off in profit, most industries will not continue support for long periods of time.

The funding of a consortium may be highly variable. One customary procedure is to charge all industrial components an annual fee which may be based on the corporate sales. In return the consortium will provide access to expertise in the university faculty, publications, non-exclusive patents, seminars, graduate students for summer or other employment, annual meetings for dissemination of information and services such as literature searches.

The consortium usually has several organizational parameters which are present regardless of its form. These include a focus of attack on a problem, possibilities for funding, ability to recruit personnel, a management structure, and formal agreements between all parties concerned. Certain conditions should also be applied to all consortia. There must be a central point of command; governance by committee is useless. The consortium must work as an institution, not as the handmaiden of an individual who may have conceived it. The consortium must satisfy mutual goals of all parties or it will not continue. The goals must be apparent and clearly stated and understood by all parties. An administrative structure must be provided OUTSIDE the structure of the university or the industry or the project cannot succeed. The consortium must have a home and agreements must spell out in detail how space and facilities will be provided.

Most consortia have an advisory board of representatives from all interested parties. Many groups have two advisory boards, one for technical input and one for university and industry representation. Most provide a specific path for technology innovation and transfer. This may include such items as publication rights (industry often wants publication delayed until exploitation is under way) and time before publication (usually 30-60 days), patent rights (industry may want exclusivity), royalty payments, etc. and these must be clearly spelled out in the agreement.

Obviously, one of the major aims of any consortium is the establishment of direct communication between the elements of the group. Without communications with complete trust and reliance, the effort cannot succeed. This is a critical problem and has resulted in the failure of many consortia because of the different aims of members.

The interest of industry and often of the government for consortium formation is the innovation and technology transfer from the research lab to practical purposes. Industry, especially large industry suffers from a lack of entrepreneurship and for it to take a risk on new products requires an unacceptable risk of net worth. As a result the company often simply wishes to improve old products. This often places the industry at odds with academia. Industry is interested in rapid transfer of knowledge, exploitation, and return of investment while the university may be interested in full development of an idea. This is a increasingly critical situation when the Germans, the Italians, and especially the Japanese are able to develop products more quickly, produce them with greater efficiency and in higher quality than American counterparts.

Consortia organized with the government as a partner may not be a perfect solution. Much of government work is done on contract with RFPs (Request for Proposal) which are very specific and do not allow deviation in research directions. However it is possible to consider the organization of consortia involving all three elements.

Researchers always believe their invention will revolutionize an area of science but all too often the better mousetrap only traps the consortium. So determination of future fair market value is critical to success. Technology can get high marks only if it has the potential to create new products that can be patented and protected, and have a high market potential. The availability of funds is critical and this depends in part upon the image. If a research consortium is a success, the output will be a saleable product. Exploitation of the idea then becomes a major concern. There are several established methods. The MIT method is to spin off a small company which produces the product, makes profit, uses faculty as management or technical staff, and returns a percentage of the profits to the institution. MIT has been phenomenally successful in this type of operation. Many universities would not be happy with faculty engaged in employment outside the university or with teaching and research time diluted by management requirements. In other cases, a university has set up an institute or similar organization within a department or the university to exploit the ideas. Industry may contract with the institution for total output or a share of the output of research and the institute is free to seek government contracts or grants. A separate freestanding organization may be set up as a nonprofit corporation which can receive grants and contracts, license industry and conduct research and development. This organization may bleed faculty members from the university as consultants or researchers. Industry may elect to be the source of support, as with SRC mentioned above, and provide facilities, employ faculty on a part time basis and direct the research toward industry goals. Each organization may be composed of a full time staff with part time input from the elements of the consortium. Examples of each kind of organization can be found.

A new type of consortium is beginning to develop as the individual states realize that development of industry and therefore state development depends upon a research base. Virginia, for example, has set up a Center for Innovative Technology which makes grants to university researchers, and focuses on technology development and patenting with transfer and commercialization of products which appear to be feasible. The cooperation with industry is stressed. The state, through the CIT, has elected to develop specific areas where major expertise exists in the universities. In four years CIT has involved more than 200 industries in 300 projects in some 20 or more institutions. Texas has developed an Applied Technology Initiative in which large grants are made to potentially commercially feasible projects in universities. The universities are expected to exploit the ideas as they develop. Other states have plans which serve the same purpose but vary in design.

The research consortium is an entity whose time has come. There are now 50 or more industrial consortia and dozens of university based or supervised organizations. All are concentrated on the technology transfer of ideas developed in basic research and all encompass academia and industry with the government playing an increasingly larger role as the regulations governing patent protection of federally supported projects loosen. Hopefully, such arrangements will again permit America to take the lead in world markets and technical development. We need only mention that VCR, television, audio recording, and many other ideas were developed in the United States but NONE of these highly valuable commercial products are produced here. Ten years ago we lead the world in production of computer chips but the Japanese now have 90 percent of the market. If the United States is to proceed from the development of ideas in which it excels to the production of a viable product in which it lags behind most of the other developed countries, consortia between industry and academia may be the best answer.

I have been instrumental in the formation and management of several consortia, working closely with NASA, DHHS, and the Indian Health service to create a consortium to provide health care to the Papago Indians (Chapter 4); with the Japanese in the creation of a consortium with Sumitomo Corporation, Baylor College of

Medicine, the Harris County Hospital District and the University of Houston to create a consortium to exploit the credit card medical record (Chapter 2); as director of the Southwest Research Consortium in San Antonio, Texas pulling together the resources of the University of Texas, the Texas Medical Center, the Southwest Research Institute, the Southwest Foundation for Research and several other institutions to create a biotechnical research base. The work has resulted in participation as keynote speaker in several symposia held at the University of Wisconsin on the formation of consortia which further sharpened ideas and methods.

As discussed in a general way above, each consortium had distinct individual problems which had to be solved before it became a success. The Papago Project, for example, required that NASA and DHHS cooperate closely in development of the idea. DHHS refused to contribute money to the project but did give in kind, though not to the extent that NASA provided. But DHHS wanted equal credit. In addition, the tribal mores of the Indians had to be considered and appeased before a system could be developed. In the Southwest Consortium, it was necessary to try, often unsuccessfully, to meld the desires of a group of basic scientists with ambitions to expand theoretical research with those of a group of engineers who wanted practical results. The Sumitomo project required the cooperation of a group of physicians interested in the provision of health care and who tolerated no interference with a group of engineers and computer specialists who understood very little about medical problems. The success of each project was conditioned by the sociological factors. As a result the Southwest Research Consortium was not a great success, the NASA project was successful, and the Sumitomo project was a partial success.

REFERENCES

Brown, J.H.U.*The Consortium in Scientific Development,* Society of Research Administrators Journal, Winter 1978 , p. 45.

Brown, J.H.U. *The Research Consortium.* Research Management, May, 1981. p. 38.

CHAPTER 9

PUBLISH OR PERISH

One of the first considerations of any scientist is publication. Unfortunately publication has become the *sine quo non* of achievement in science. For most scientists the final aim of research is publication and recognition. As a result the syndrome of "publish or perish" becomes more prevalent every year. The pressure for publication is clear. Promotion, tenure, access to other scientists, and reputation depend upon the number and quality of publications. As the number of scientists increases, the problem becomes more acute. We must devise a system that will control publications in both number and kind before we are swamped in our own paper. The scientific community needs to rethink the entire question of policy for promotion, tenure, publications, authorship and ethics all of which are closely linked in a pattern that began with the first scientific investigation and has escalated since. Unless scientists are willing to be more closely regulated in what they write and how they write it, public confidence is likely to be lost. Confidence as measured by the public press appears to be waning now with the repeated charges arising from first rank universities about cheating, publishing methods, and failure to control reporting of premature results. Problems arise in terms of volume of publications, numbers of authors, and laxity in the research itself.

A publication of the Council of Biology Editors estimates that one million scientific papers are produced each year by the two million scientists in the world. On average, each scientist in his/her life time will produce about ten papers. This mountain of material is published in some 75,000 journals in many languages. Maurice Visscher, professor at the University of Minnesota, estimated if a scientist read ten papers per day published in his field, at the end of a year he/she would be two years behind in reading. And the number of journals is increasing by more than 10 percent per year while the cost of each issue is increasing exponentially. As a result the number of subscriptions per journal is decreasing and the cost per issue is increasing, creating a vicious cycle because journals must raise prices as subscriptions decrease and costs increase in order to continue publishing.

The publication problem is not likely to change rapidly. The requirement still holds that a paper be published in a "reputable" journal in order to establish priority and obtain promotion and tenure. Thus the proliferation of journals is not surprising when it is considered that scientists believe that eighty percent of all consideration for promotion and tenure is based solely on number of publications. This is borne out by surveys of promotion procedures of large universities.

Because success in science depends largely upon the number of one's publications, the number of authors per publication has increased astronomically. The number of authors on some papers has been as many as seventy five in some areas of physics. There is often a serious question of the contribution of each author to the total paper. A. Relman, editor of the *New England Journal of Medicine*, found that he was forced to restrict the number of authors to be included on any paper published in that journal unless the role of each is carefully designated.

In part, because of these attempts to increase publication rates by individual scientists, measurements of publications have been facetiously proposed. The Least Publishable Unit (LPU) has been suggested as the minimal amount of data which can be produced and still be published. This allows the breaking of a research effort into many portions, each comprising a separate paper. As a corollary, the Maximum Recycling Quotient (MRQ) is suggested as the number of times a given set of data can be used in succeeding papers or in several journals without arousing suspicion of the editors.

Formation of new journals is encouraged by the average length of time required for publication which may be more than a year in some cases and is rarely less than six months, and by the rejection rate of articles submitted to well recognized journals which may approach eighty five percent. Ziman has remarked that it is not clear whether a new journal is designed to publish worthwhile results or to relieve the pressure by publishing articles which are rejected by the more prestigious journals. This observation is borne out by the studies of the Rockefeller Foundation which found that most articles are eventually published often after repeated rejections by the better journals.

The thrust for rapid publication has occurred because it has been claimed that seventeen percent of scientists are scooped by other scientists in the same field and rapid publication is therefore essential. The "hot" discovery and the quick announcement are routine and some journals are now designed for quick publication solely to establish priority with the aim of publication at a later date. This practice raises questions of validity of data, publication without review, and risk of publishing data which cannot be confirmed. Some articles in genetics have been published in 10 days and it is not uncommon to see publication within a week or so. Obviously the lengthy approval and publication process is not solely to blame. Some publishers and scientists also fail to handle properly and review articles. *Science* in September 1980, published an article by Peters and Cec in which they took ten articles from prestigious journals and resubmitted them to the same journals with different author's names. Only eighteen percent were accepted the second time around. Of greater import, only ten percent were recognized as having been previously published. This raises serious question of the quality of review.

The usefulness of any publication can be questioned. Garfield analyzed papers in physics and found that fifty percent of all important papers as measured by the number of times they were cited by peers were published in twenty three journals and ninety percent were in twelve journals despite the fact that hundreds of physics journals are published. One may argue about the worth of citations as a measure of quality but if papers are not used by the scientists, they can hardly advance the field.

Garfield attempted to judge the qualities of both journals and research as an offshoot of his publication, *Current Contents*. His system was simple. If a journal is well known by scientists in the field, the number of papers cited from that journal by other than the author should be a measure of the research importance. Garfield compiled a list of journals and citations and found that the agreement was excellent. The *American Journal of Physics,* the *Journal of the American Chemical Society, Science, American Journal of Physiology* were among the most cited. He found that some 55 journals contain ninety percent of all citations. This again points up the fact that most articles are published in relatively unknown journals and are not cited by others. The question then occurs: do publications serve as an ego salve and a promotion scheme for scientists but do not contribute to the advance of the field?

If it is possible to rank journals by scientific citations, it is also possible to rank scientists by the same criteria. If a scientist is cited by others, his/her work has some relevance, if not, the work may not be worthwhile to have been published or supported through grants. Recent surveys suggest that an average of 55 percent of *all* work is never cited, and that in the major scientific areas such as chemistry and physics 35 percent is not cited. In a short if not decisive test, I counted my own citations in the Index. As I could have predicted in advance, some articles were not cited at all, and some were cited one hundred times or more.

There is another way to look at the problem. The question of importance of any publication to the development of a field or to science as a whole is critical. In all of the scientific literature only 5000 papers have been cited more than 2000 times; only 50,000 have been cited more than one hundred times while some 12 million papers have been cited only once or not at all. In total about six percent of the journals and scientists contribute seventy five percent of the citations. This has lead to proposals from legislative bodies that support of research should be limited to the top group of scientists for they will contribute most to science and make the most discoveries.

As would be expected, Garfield found that the more prestigious the scientist, the more he/she was cited-an award such as the Nobel Prize increased the citation rate markedly. Conversely, the fact that the individuals most cited are also most likely to win the Nobel Prize later establishes credence for the survey method. More than fifty percent of *all* scientists are never cited while members of the National Academy of Science are cited on an average 70 times. Only ten percent of cited papers come from undistinguished departments (as selected by reports of colleagues). The data support a conclusion that much of data collected is unused. This does not negate the archival necessity and interest.

It is also possible to determine the most important (?) science departments by the same set of criteria. Garfield found that sixty percent of cited work in a field came from ten distinguished departments in universities. These departments were also recognized as important by the National Research Council Educational Survey conducted by the NSF at regular intervals and by colleagues. Rankings are also published yearly by the *U.S. News and World Report.*

Despite the rush for publication, the number of papers published by an individual is no sure criteria for fame and fortune. Einstein won the Nobel Prize with two papers and has been cited thousands of times. One scientist has published 650 papers and many have never been cited. He did not win the prize. Another published a paper every ten days during his active life. Citation analysts also take into account the fact that in a very small field, or one which is developing, few citations may be expected, and they may be published in obscure journals. An uncited article may provide a basis of future experimentation and a very bad article may draw many citations of rebuttal which do not contribute to the advance of science.

Nevertheless, the citation index has been used for determination of promotion and tenure. The *Chronicle of Higher Education* has reported legal suits alleging that persons with fewer citations were promoted over persons with greater numbers of publications, indicating that the citation method is considered to have some validity. Many departments list citations as a measure of departmental prominence.

Some $12 billion per year are spent on publication with an average journal cost now in the hundred dollars per year range. After publication an article requires abstracting, library services, time to find information, computer time and programming and review which makes the total cost astronomical. As a result, some attempts are being made to determine the worthiness of any paper. Review is not sufficient. A journal with few readers in a highly specialized field may take many articles which would be rejected by a more prestigious publication; length is not a criterion; number of authors may be a detriment; field of research may be important; and many additional parameters could be cited but no definitive proposal for control has yet been made.

The present scientific environment which demands more publications leads to cheating. Pressure of time and competition with colleagues tends to result in hasty decisions, not well thought out experiments and rapid and multiple publication. Forging of data, reuse of data, and inventing data are becoming more and more common. The practice is not a new one. Merton points out that in 1830 Babbage, the famous mathematician, wrote on fraud in science and divided the schemers into "cookers" who take many observations, select the ones which fit the hypothesis and publish the results and "trimmers" who eliminate the outer limits of data and publish only the consistent parts. The habit does not encompass only poor scientists. Millikan in his Nobel Prize work is said to have eliminated forty percent of his data. Mendel could not have obtained the consistency of his genetic data by chance. The famous "case of the midwife toad" where out and out fraud existed as documented by Koestler is another example.

Most of the scientists who are familiar with their field are aware of one or more of such practices. There have been times I have been guilty of reworking an article for a journal with a different slant on request of the editor. In my work with the laser card (Chapter 2) I have used the same data to publish an article on the technical aspects of the equipment and another on the results of the clinical trials. One appeared in an engineering and the other in a medical journal.

Petersdorf lists unethical behavior of authors as unidentified authorship, dishonest data, and duplicate publications. As an extreme, in the famous Darsee case of falsified data NO author was willing to take responsibility for the paper. In at least one case, the first an individual knew of his name listed on a paper was a request for a copyright assignment. Some authors have reported that their names were used without permission and it is common practice that the head of a department heads the list of authors regardless of the extent of his/her contribution to the work done in the laboratory. In my own work, if my contribution was equal to or greater than that of the other authors my name has been listed, but it was never listed where a real contribution was not made to the research.

In many cases there may be an honest mistake caused by failure to observe good scientific practice. Bailar has listed these as errors of data collection, creating *posthoc* hypotheses, misuse of data or wrong interpretation, poor statistical methods, selection of data and adjusting the data to fit circumstances. Editors have recently added other criteria. The "curve fitter" produces a curve from two points, the "milker" has long discussions with little data to back it up, and the "conjurer" attempts to prove his point with statistics rather than data; while the "generalizer" attempts to makes broad conclusions from little data.

These practices have resulted in Daniel Tosteson, Dean of Harvard Medical School, announcing a policy of quality as opposed to quantity. Harvard will consider only five papers in a promotion of an assistant professor, seven papers for an associate professor and ten papers for a full professor regardless of the total number of publications. In order to help prevent the rash of multiple authorships, Harvard also will consider papers only where "intellectual contribution" can be proven. This may prove to be a disadvantage to the scientist. It is clear, for example, that the number and size of grants made by NIH is directly correlated (at the 95 percent confidence level) with number of publications of the grant recipient.

The situation is spreading to the young. Surveys of medical students reveal that eighty eight percent have cheated at one time or another. There is no question in my mind but that cheating has increased in medical classes. Students entering medical school no longer have the same moral responsibility of the students of forty years ago. However, a fellow student in graduate school with me passed many of his examinations by removing the stencils for the exams from the waste containers and reproducing them.

Science must assume a responsibility for assessing worth of a research project, deciding how the results should be published and defining the authorship(s) of any paper through clearly developed guidelines clearly stated for all scientists. Otherwise we will continue to be inundated in a sea of paper.

A revision of the publication system is needed. Two lines of attack are indicated. New and cheaper methods of publication must be devised. A more selective method of reviewing papers must be arranged and the dependence of promotion and tenure upon numbers of publications should be changed.

Regardless of the sociological basis of publication, each scientist has to approach the problem of producing articles on his /her own. The ability to write is critical. One may experiment, collect data and reach opinions about its importance, but unless the material can be put down in a readable, clear and forthright manner, the effort is wasted. No one will plow through notebooks to find the concealed information. In some thirty books and more than one hundred and twenty articles I have tried to write clearly and objectively.

The question is often asked, "How do you write a book and several articles a year?" The answer is simple. One works at it. Writing a book occurs in stages which are clearly delineated. The first is the idea. The right idea may take months. It usually begins as a hint which slowly becomes filled out with details over time. The idea is never the complete book. It is a statement of what the book will be and something about the contents.

To flesh out the idea into a form that is useful, reading is necessary. Days must be spent in the library reading everything that may have a bearing on the idea. Much will be later discarded but everything that may fit into the pattern must be examined. With enough reading, the idea will begin to shape itself into a substantial form as the additional materials are used to fill out the details.

Once material has accumulated, an outline which pulls the idea into shape and fits the reading material into the concept must be prepared. This may take a simple form of a series of statements of what is to be accomplished or it may be an actual outline of chapters. The book is now beginning to take shape.

Writing or rather the intention to begin writing is a critical part of any book. One must to be prepared to take much time at first. Days go by with only a few words in the word processor. Ideas do not seem to gel. References are chaotic and do not fit the ideas. Suddenly, everything clicks and the words begin to flow. Ideas are fleshed out into substantive statements. Writing then becomes easy. But it is not unusual to take six months to think about the contents of a book and only three months to write it.

The chapters must be thought out in advance. Each chapter has a theme and the material in that theme is included with references to materials to be used. Although each chapter is on a different subject, it is important to keep in mind the thesis of the book as a whole and be sure than each chapter fits into the whole. I have often written a good chapter only to discard it later because it did not contribute to the overall approach.

Now it is time to consider the book as a whole. The chapters are arranged and rearranged until they occur in an appropriate order, the headings are devised or altered to fit the general theme and the book is almost finished. Chapters may still need to be subdivided or combined. Subheads may be necessary to divide ideas so long as they fit the general theme of the book.

Details have to be finished. References need to be checked not only at the source but between chapters if they have been used more than once. An index must be prepared and the bibliography assembled. The index is a difficult part of the book for an amateur. Every item which might be referred to must be separately cataloged on a separate card, the cards must be alphabetically arranged and the cross references must be established. For example, if an item appears as a disease, it also may be listed under symptoms, diagnosis, treatment or other categories. All possible combinations must be tested.

The Introduction is often the first part of any book to be written because it explains what is to be done and why the book was written. A preface must be completed. Many say that a preface should be done before the book is started, but a preface may be written at the beginning or the end, however the fancy strikes. The introduction and the preface are two totally different parts of the book. The preface usually tells why the book was written, problems to be dealt with in reading the book, contributions of others and other general matter. The introduction outlines the book, establishes the philosophy under which the book was written and provides the lead in to the individual chapters.

The preparation of the book for the publisher is critical. Most books are completed on word processors. The punctuation, spelling, sentence structure and other details of the mechanics of writing must be meticulously observed. The publisher usually provides a format but in general the manuscript must be double or triple spaced, wide margins must be used on 8 1/2 x 11 paper, and the type must be clear and 12 point in size.

Finally, the book is typed, assembled and ready to mail to a publisher. If one is lucky and has assurance of publication, the satisfaction is greater but if there is no publisher on the doorstep, one must be found. Libraries have reference lists of publishers with names and addresses. I try to pick one who is publishing in the area in which I am writing. The best source may be the book "LMP", *Literary Market Place,* which lists all publishers and the areas in which they publish.

Some authors swear by the use of agents to locate publishers. Again the *LMP* provides lists of agents and their areas of expertise. I have never used one. My books have been solicited in advance by a publishing house, or I have mailed them directly to a publisher. The first try is not always a success. It is sometimes necessary to try several publishers before one is found who is interested in the project. But I have never failed to find a publisher among the great houses.

Once a book is accepted a contract is forwarded to the author. The contract specifies the duties of the author and at first reading usually seems that the total responsibility is that of the author and the publisher assumes no risks. This is not the case-if the book does not sell the author receives no royalty but the publisher may take a large loss because of time and money invested in editing, printing and distributing the book. The contract may specify the royalty to be paid to the author. Under normal circumstances with an unknown writer, the royalty may be 10 per cent of the book sale prices for the first few thousand copies and it gradually increases as sales increase. News of advances of hundreds of thousands of dollars are commonly seen in the newspapers for fiction writers of known reputation but an advance is not common for most nonfiction books and in any case is likely to be small.

Now the waiting begins. Months go by with occasional queries about the book from the person assigned to be the editor. Finally, proofs arrive. Originally, there were two sets of proofs. The gallery proof consisted of long pages with a column of print on each page. It arrived with some editorial corrections and the author was expected to make additional corrections of spelling, punctuation, alteration in text and any other changes directly on the gallery which was then returned to the publisher. The author was encouraged to make few changes in the text although alterations were possible. The corrections were made in the type and format and the book was then printed as it would appear on individual pages and the page proofs were again returned to the author. Again changes were permitted but any changes that were not due to the typesetter's errors were charged to the author. The page proofs were returned with corrections and the book was now underway for final printing. At this point the page references for the index were prepared and carefully checked against the proofs. But until the final paging is determined it is impossible to match index items against the page number of the page on which they occur. Nowadays in the hurry for publication, many publishers skip the galley proof and send only page proofs to the author. If no major changes are made in the manuscript at this point, indexing can proceed.

The present modes of publishing which now use electronic devices for typesetting and photography for final preparation operate in another manner. The author may be asked to provide a camera ready book. This means that the book is typed exactly as it will appear when published using a word processor to set the type format and size (Times and 10 point are satisfactory), and the size of each page must be matched exactly to the final book size (4.5 x 7.25 inches is a common book size printing out as a 6 x 9 inch book). Indexes, chapter headings, table of contents, and other elements must be set up exactly as they will appear in final. The publisher then photographs each page and prepares the printing mat for a rotary press. This type of preparation requires that each chapter head begin on the right hand page and that chapters preferably end on the left hand page. It is not unusual for the publisher to ask the author to furnish a computer disk with the book correctly formatted on it. This can be fed directly into the printing process.

In the production of a book by the galley proof method, the author has a chance to make many corrections as he/she proceeds through several stages of preparation. If only page proofs are submitted to the author by the publisher, the chances of major corrections are not great but may still be made. The camera ready book has none of these prerogatives. No errors are permitted.

The book is now totally in the hands of the publisher. The publication date may be months away. The speed of publication depends upon the stream of books being printed into which an individual book must fit. If the author has taken a long time to make corrections or return proofs, or if much editorial work is required to ready the book for publication, time delays may be very long. The author may or may not be consulted about the dust cover illustrations and format and the color and type of binding. Finally the author receives copies and the book appears on the bookstore shelves. The next wait is for six to twelve months before the royalty check appears.

The publication of a scientific paper is very different from book publication. The book starts with an idea and data is filled into it from reference material and available data. The paper starts with the data. Preceding the data is, of course, the experimental design, clearly definable results, a sharp conclusion and discussion.

Then a journal must be chosen. There are two reasons for making a very careful choice. One is the need to select a journal publishing papers in the general area of the research. Failure to pick a proper journal will result in almost certain rejection. Often the choice may be very narrow. Should one publish in the *American Journal of Physiology* or in the *American Journal of Applied Physiology?* The selection may determine whether an article is accepted and whether extensive rewriting will be required. Should one assume the results are so important that they must be published at once and therefore should be in *Science* or other frequently published journal, or is the article to be definitive and complete in which case a journal with a longer publishing time which will publish articles with more detail and analysis be considered? Finally, the journal must be examined in detail. Each journal has a definite format for presentation of material including specifications about data, conclusions, references, and layout. Each journal publishes such details at least once a year. All must be followed.

Write clearly. The turgid texts of most articles, the lengthy explanations and the circumlocutions of most scientific explanations beggar description. Simple sentences, clear exposition, and simple straightforward reasoning are essential. There is a scientific jargon which obscures the facts. Avoid it.

As a rule, the article can be submitted to only one journal at a time. Most journals require a signed statement that the article will not be submitted to any other publication while it is still under review. The journal will usually send the article out to be reviewed by one or more editors or reviewers. Each reviewer may reject the articles, accept it, or may require changes in the material. In any case it is permissible to inquire the reason for changes, to ask for additional reviewers, or to submit additional material. If the article is finally rejected, the best plan is to take the suggestions of the reviewers, incorporate them into the article, rewrite the article in the format of another related journal and begin the process again. By sending an article to as many as four journals I have not yet failed to publish.

As a member of the editorial board of several journals, I have been struck by the poor writing in most articles. The experimental procedures are not clearly stated, the data are not presented in tables or graphs which can be easily interpreted, the conclusions are not clearly expressed or are not substantiated by the evidence and credit is not fully given to others. This may well explain the high rejection rate of most popular journals.

The format for a review article lies somewhere between the paper and the book. It must researched much like a book, the ideas not the experiments are critical, but the format provided by the editor for references and text must be followed and the length may be specified. The reasoning must be critical and sharp because the article is usually on a narrow topic which must be reviewed clearly and definitively. Articles on the subject must be examined both positively and negatively. Some reviewers take a position on the correctness of the data while others simply present both sides of a question and allow the reader to make a choice.

The job of an editor of a series of books compounds (confounds?) the process. The usual jobs of an author must be assumed but in addition the editor is required to locate the authors of specific chapters, cajole them into writing and force them to produce the commited article. And this is no mean job. Trying to persuade a reluctant author to produce a manuscript in a given format and with a definite deadline can be deadly. When manuscripts finally arrive, the editor has his major task-revision, elimination, forcing all papers into a common format to produce a book which looks finished and correct and satisfying authors about changes made by the editor. I have been sued by an author for removing pages of computer printout from his manuscript which contributed nothing to the argument and forced to rewrite entire articles because of mistakes in grammar. A major task is producing or reproducing the figures and graphs in the same format. In addition, the royalties are poor. The editor receives about fifty percent while the authors split the remaining fifty percent of the royalty. There is one exception. The rapid publication of a symposium is not as difficult as producing a finished book. Papers can be accepted and photographed as is. This makes for a rapidly produced volume but one which is physically poor in that type faces may vary, different styles may be used and the jargon may vary from author to author.

Finally, the scientist may be asked to write a public information article for a magazine like *Scientific American* or *Bioscience.* Such articles are most difficult to write. They must be scientifically accurate but they must explain phenomena which may be complex in a clear and open fashion for the semi-lay reader. To maintain objectivity and still enthuse about events in science is hard to carry off.

The modern scientist must take another interest in publication. It is essential that the scientist be prepared to write letters to editors of newspapers, write newspaper articles or respond to questions from the press regarding the political or social aspects of science. The public with an axe to grind, such as the antivivisectionists, do not hesitate to use the media to advance their cause. The scientist must reply in kind or take the consequences.

Despite the problems of publication for science in general and for the individual scientist in particular, the joys of publication exceed the sorrows. There are few joys which are greater than opening a package from a publisher and finding pristine new books with the unmistakable smell of printers ink and one's name on the cover.

REFERENCES

Wood, J.L.. *Scientific and technical journal proliferation.* CBE Views 4: 4 (1981)

Angela, M. and A. Relman. *Redundant publication.* New Eng. J.Med. 320: 1212 (1989).

Ziman, J.M.. *Information, communication and knowledge.* Nature 224: 318 (1967)

Garfield, E. *Citation indexing.* New York, John Wiley and Sons, 1979

PART II

TREADING IN ACADEMIA

CHAPTER 10

IN THE BEGINNING

I was not born a scientist. In fact my first real exposure to any science was in college. There was little or no science in a small west Texas town with an inadequate school system. In those days the person most likely to teach science was the football coach who was charged with teaching hygiene. As he usually had taken the minimum science necessary to graduate in physical education from his college, there was likely to be little interest in and little knowledge of any science.

In one way I was fortunate. At an early age, a doctor misdiagnosed me as having rheumatic fever and put me to bed for several months. The time was spent in reading beginning with Tom Swift and the Alger series, on to western novels of Zane Grey, and then into anything which could be found. I obtained a quirky, mismanaged and spotty education with an eclectic choice of reading matter.

High school was completed in Stockdale, Texas. The only teacher I remember was Sara Bell Eardley who came, newly graduated from college, to teach English. She had a spectacular figure and she wore stripped dresses which emphasized it. I would have married her there and then. She did teach me English, some poetry which I can still recite, and the rudiments of grammar.

Mostly it was a desert of learning. There was no science and the library consisted of a small alcove in the back of the auditorium The library was open at widely scattered times but I found that if one released the catch on the window in advance, it was possible to return late in the afternoon or on weekends and sneak into the school. Dr. Eliot's Five Foot Shelf of Books with particular attention to sex detailed in the "Affairs of Cellini", an outdated encyclopedia, and some history were read sitting crosslegged on the floor to avoid detection. There was nothing of interest in science and I graduated without any information in the hard sciences.

I graduated with honors from Stockdale High and received a scholarship to Southwest Texas State University. The University in San Marcos was small, with about 1500 students and poor in both money and quality. I had thought often about the University of Texas which loomed over the Texas academic horizon but my father wanted me to start to work and forget about college. Only the scholarship and a job I obtained in San Marcos made it possible for me to pay most of my own expenses and go to college.

In 1936, the world was just beginning to recover from the depression although it would not recover completely until the Second World War. The Federal government had established the NYA (National Youth Administration) to employ young men (women were not considered in those days) and I worked weeding flower beds, mowing lawns, and helping in construction around the college and later as the custodian of a men's dormitory and so obtained free room and board. Work in the library as a filer and desk clerk, and in the dormitory as a waiter and short order cook was also available.

I also made money in a slightly unofficial way. My family moved to San Marcos and my mother took in coed roomers to make ends meet. I often wrote themes and one case a thesis in education for the girls. Although there was no official pay schedule, the idea was conveyed that a reward would be expected if grades were satisfactory. They usually were and I was usually paid.

Despite a full work schedule, I took overloads each semester and graduated in three years. Many of my teachers come to mind. Dr. H. Greene, taught history with elan. He would have been the iconoclast of today, slouching in unpressed trousers, and suspenders, his gray hair unusually uncut and uncombed, as he peered myopically over round glasses. He chewed tobacco in class and spat in his desk drawer. One remark made him famous on the campus, "I've tried nine methods of birth control and have a living example of each one" This was considered the height of daring in an age where sex was not openly discussed. There was the physics teacher, Mr. Rush. fat, bald and unatractive but I heard that when he died two families, each claiming him as the head came to the funeral.

Dr. H. Keys taught chemistry and involved me in my first real research project. It had been noticed that salt cedar which covered thousands of square miles of Texas had an aromatic odor and a high rate of sap production. We ground up whole trees, extracted the sap and distilled it to produce a variety of compounds. When the separate fractions were analyzed, one was determined to be a derivative of camphor and we thought of riches when it was discovered that Kodak used tons of it in film processing and the supply of camphor from the Far East had been cut off by Japan. Unfortunately, by the time we learned this and thought about a pilot plant, synthetic camphor had taken the market. But I can still remember the acrid smell of cedar resin as we ground trees to a pulp and processed them in vat lots.

Dr. Syrell Wilkes was a professor at San Marcos with a shiny new Ph.D. from the University of Texas. He arrived on the scene determined to make a showing in biology under the professor and chairman, C.S. Smith, who had long passed his prime. Wilkes was handsome, with blue eyes and wavy blond hair and rapidly became the target of every coed on campus. One caught him. Under his encouragement I made a survey of the flora and fauna of the natural springs in San Marcos which was published in an obscure journal. It made me locally famous, my name was mentioned on a local radio program. We also became interested in the caves of the Hill Country of Texas. These limestone caves often began as a sheer drop of fifty or so feet, then would wander through the mountains for miles. There were stalactites and stalagmites, underground rivers, millions of bats, and a feature unique to these caverns, the blind colorless salamander. In a later time, I attempted to find the thyroid gland in these creatures living in a constant environment by using radioactive iodine and trying to localize its accumulation. We could not find one. I still do not know if the salamanders have any form of thyroid tissue.

Iben Browning, Santry Greene, Robert Lowman and I explored some of the caverns. One trip almost became a tragedy. We had explored a cave which was entered by a fifty foot vertical drop. We were attempting to take soil samples to determine the bacterial content in deep caves. After an all day exploration deep in the cavern all of us except Iben had reached the surface when a rock dislodged and hit him in the head. There were no safety helmets in those days and he was knocked unconscious. He fell on a ledge about halfway down the hole. We watched in horror as he wakened and tossed in partial consciousness. There was no way one could reach him and no way one could bring him to safety. We called and threw small pieces of gravel to attract his attention. He finally roused, realized the situation and sat quietly for several minutes before he could move and clamber up the rope to safety. There was a universal sigh of relief. Iben went on to become a protector of the environment and achieved notoriety when he predicted an earthquake in the central United States which did not occur.

At San Marcos, I found that I had a bent for sciences. My father insisted on my taking a business major but the curriculum could not arouse my interest. Good grades in all sciences and the little research mentioned above attracted me to the sciences. During the school year I appropriated a room in the physics area which converted into an office of sorts. Accumulations of books filled the little room until the head of the library called me into his office and threatened disciplinary action unless they were returned.

In the three years at San Marcos, I can recall no dates and almost no contact with girls outside the theater. It seems in retrospect that my life was mostly work and school. But it could not have been not all work I acted and built sets in the theater and was elected to the theater honor society. I debated for the college and was again recognized by election to the national forensic society, Phi Kappa Delta. I made the honor role with election to a national honor society. Finally I graduated in 1939 second or third in the class.

My academic career was determined at that moment. Wilkes recommended that I go to graduate school at the University of Texas in Austin and his recommendation brought me a small stipend. Against my fathers wishes I accepted and moved to Austin in the Fall of 1939.

CHAPTER 11

ACADEMIA FROM LOW ON THE TOTEM POLE

The University of Texas in Austin had a distinguished department of biology. The geneticists were well known and Thomas Painter would win the Nobel Prize. The department of physiology was headed by E.J. Lund, Swedish, a tall, heavy set man with a shock of wavy gray hair and a domineering manner. Lund was an authority in electrophysiology. Although the science was being used daily in medicine (the ECG and EEG), Lund was trying to measure electric potentials in plants and correlate them with growth, effect of light and other parameters. He had a large number of graduate students. One was John Paul Stapp, who later went on to test the high velocity sled to determine the effects of acceleration for the Air Force.

My introduction to the department was not auspicious. I had lived in a small town, gone to a small college, and was now thrown into a large university, into a research milieu and into contact with a much more sophisticated research environment. Lund had offered me a position as laboratory assistant and I began by teaching Zoo. 6, the beginning biology course. He then introduced me to research in his usual brusque way. I was told to make cytochrome C. I had no idea of where to begin and he was not about to enlighten me. After much reading in the chemistry library I found that the extraction had been done in Germany and was published in the great German chemistry manual, Beilstein. I had a smattering of German and I spent weeks in deciphering the chemistry to prepare the compound from Lund's favorite tissue, the onion root. I obtained a substance but I had no way to determine if it were the actual compound-I had never been introduced to spectroscopy.

The next year Lund had me take over his graduate course in Physiology as the lab instructor. This was an enormous task and I was ill prepared for it. He required fresh tissue for each experiment so I grew Spirostomum, trapped snails, caught frogs, and made solutions each week. We did a great variety of experiments in general physiology, many of which I still recall with pangs of anxiety.

Lund was a hard but fair taskmaster. He had a definite set of rules. For example, graduate students were required to learn such chores as glass blowing and machine shop work to make their equipment. Lund used onion roots in most of his electrophysiology experiments because they were small, grew rapidly, responded to light and gave excellent electrical potentials. Being small, the roots were also fragile and great care was needed to blow glass vials to contain them-hence the lessons in glassblowing.

Lund also foresaw the coming of molecular physiology. We were required to take biochemistry and chemistry of colloids when these sciences were totally separate from biology of any sort. The biochemistry was taught by Roger Williams, the discoverer of pantothenic acid. Williams mumbled, wandered in the subject, and was a poor teacher. The halls of the chemistry building stunk with a persistent smell of cooking spinach from which the vitamin was made. Between Williams and the milieu we learned biochemistry with difficulty.

At the end of two years in graduate school my funds were exhausted. The pay of a graduate assistant did not cover all expenses and Lund could not or would not increase stipends. So I dropped out of school and accepted a position teaching biology in the Seguin High School for which I was not well suited. The only good to come out of my stay in Seguin was that I met my future wife, Jessie Schulz. We were married in 1943.

Teaching in high school started in September and in December, 1941 we went to war with Japan. With the likely coming of war and the beginning of the draft I had applied to several government offices. In early January, a telegram came ordering me to report to Scott Field, Illinois for work with the Army Air Corps.

With a major in physics and chemistry as well as biology in college the Air Corps assigned me to study the new ranging and detection device, radar. There was no separate Air Force, it was a part of the Army and we used Signal Corps equipment installed in planes and on the ground. We went to school ten hours a day for many months receiving training in radioengineering, design, installation, testing, and maintenance of radar. I finished first in the class. We were then ordered to report to Tomah, Wisconsin, where a school had been set up train army personnel in all aspects of radio location. For more than a year we trained students in maintenance, installation, and operation of the system.

My association with the Army was never happy. It was difficult to understand rigid military discipline and I was once fined a month's pay for disagreeing with an officer. Fortunately, I received a letter from Wilkes at Southwest Texas State University, my alma mater, offering me the position of Instructor of Physics, teaching electricity and magnetism to students who were for the most part air cadets in officer's training school. It came at an opportune time. We were living in tents in the field installing equipment, and it was chilly and raining. I was more than happy to resign from the Army.

Jessie and I were married just before going to Wisconsin, and we now returned to Texas and teaching for two semesters before the draft board called. In a series of conversations, the Navy offered me a commission as Lt. Commander to teach radar at the Naval Academy. Color blindness caused me to fail the physical examination for officers. The Navy then drafted me. I served several months as an ordinary seaman and was then discharged for physical reasons. An offer came of a position in the next medical class in the University of Texas Medical Center in Galveston and the examining board of physicians considered this a worthwhile exchange for a very mediocre seaman.

As it happened, Galveston was not in my agenda. The funds were not available and the present day loan and guarantee systems were not in place. Suddenly, the problems were resolved. A telegram from Rutgers University offered the positions of Instructor in Physical Chemistry, Associate of the Bureau of Biological Research, and graduate student at Rutgers. The position in chemistry was for only one semester after which I would be officially entered in graduate school.

Rutgers was one of the oldest universities in the nation, set on a hill above the Raritan River in New Brunswick, New Jersey. It glowed with 18th century buildings, and 19th century red brick. It was a large urban university with nationally known scholars and departments, but the Biology department was located in an 1890's building that had not been renovated.

On reporting to van der Meulen, head of Chemistry, a class in physical chemistry was assigned. The class was composed largely of military officers in training and premedical students. I taught both the lecture and the lab. It was almost a fiasco. I had not taken physical chemistry for several years and the last course at the University of Texas had been highly specialized in colloids. It required immense labor to keep up with the lectures. Efforts were almost stymied by the needs of the laboratory for finding equipment, writing up the experiments, and trying them out in advance of each week's class. Being unfamiliar with many of the experiments it was necessary to devise the methods and equipment. Equipment was hard to find and was often unworkable. Somehow I struggled through the semester.

On the other hand, the Bureau of Biological Research was a delight. The chief, J.B. Allison was a gentle, kind man and the second in command, J.H. Leathem, was a noted endocrinologist. The Bureau occupied an old but well equipped building behind the biology building.

The Bureau was my first experience with industrial research. Allison was a basic scientist interested in nitrogen metabolism. With help from the graduate students he perfected the Nitrogen Balance Index (NBI) a means of determining the retention of nitrogen in the body and affording a means of studying metabolism of individual amino acids or proteins. Allison worked closely with the scientists in the research labs of Merck, Sharpe and Dohme and Squibb on products that contained nitrogen and could have commercial value. For, example, we tested AMIGEN, the first protein hydrolysate to be used in clinical treatment of debilitated patients.

One of our first tasks was to investigate composition of life raft rations for the Navy. The rations had to be compact, furnish energy, contain proper nutrition and be tasty. We tried variations of rations with different amounts of carbohydrate, fat and protein, testing them on dogs to determine the minimum excretion of nitrogen and the best combination of materials. It was a chore. As we varied the composition of the diet that was molded into small bars, it changed from chalky to greasy, from sweet to tasteless. We finally hit upon a composition that used sugars other than sucrose and dextrose, had starch rather than sugar, and was lower in fat. We tested the diet upon dogs and ourselves. But we never knew if the ration was used by shipwrecked sailors on life rafts.

We used the dog as an experimental animal. We had a large kennel and a group of dogs who remained with the lab for years. One Sunday morning a Dalmatian, Smokey escaped from the lab. Chasing the dog down the street, white lab coat waving, past the First Dutch Reform Church just as the parishioners including my wife and her choir came out the front door has been unforgettable. I caught the dog and returned past the church with her in my arms. My wife looked the other way.

Working with Jim Leathem, who had an international reputation in the fields of steroids and protein hormones taught me much about research. I worked on the assay of pituitary and other hormones, on the assay of sex hormones and pregnancy and with Allison on the metabolism of protein. My dissertation on the toxicity of methyl groups in proteins attracted some attention and resulted in a major publication in the *Journal of the Society for Experimental Biology and Medicine*. During my four years at Rutgers I wrote some ten papers with Allison and Leathem. The work paid off because I received the Gerard Swope Award of the General Electric Company for two years. It was the first award made to a biologist. Election to membership in the New York Academy of Sciences and to the Society for Experimental Biology and Medicine was an added bonus as both required demonstration of ability in research.

It was not all easy going. A highly malignant testicular tumor required the removal of one testes. There was no chemotherapy in those days so a series of debilitating x-ray treatments that interfered with my research was administered. Jessie became pregnant. She had been working as Assistant Dean of Women as New Jersey College for Women and was the Children's Choir Director at the First Dutch Reform Church, the largest church in New Brunswick. She was extremely unhappy about the pregnancy and became even more so when the child was born with hydrocephalic disease and could not survive. She took the child home to her mother. The child died shortly.

Finishing the degree, taking the final orals, writing a dissertation and managing to avoid financial disaster over the next few months while worrying about the malignancy took a great toll but I finished the research, had the dissertation accepted and obtained the degree. There is more than a suspicion that the advisory committee took some pity on me.

But the situation was desperate. When an offer for more money than expected came from Mellon Institute in Pittsburgh, an acceptance was certain. Required to report immediately I accepted and the Ph.D. degree was taken *in absentia* in1948. My work at Mellon Institute has been described in Chapter 6. When offered an Instructorship at the University of Pittsburgh I accepted and taught graduate courses in biology. It was my first real teaching experience and I quickly learned about the sociology of teaching. In my classes in advanced physiology were several graduate students from the laboratory. One, Scric, made "C's" and was greatly incensed. "I'm your graduate student and you should pass me." he argued. I didn't and he didn't. It was my first experience but not the last in attempted academic blackmail.

REFERENCES

Allison, J.B., J.A. Anderson, and J.H.U. Brown, *Evaluation of Proteins in Hypoproteineimic Dogs*. Journal of Nutrition, 31: 327 (1948)

Brown, J.H.U. and J.B. Allison. *Effects on Dietary Methionine and Arginine on Adult Rats*. Proceedings of the Society for Experimental Biology and Medicine 71: 812. (1949)

CHAPTER 12

PROFESSOR

One of the criteria for success in science should be that each step requires a demonstration of ability. Unfortunately such is not always the case. In recent years there has been a relaxation of standards. Certain minority groups have obtained promotion because of race or color rather than ability. Fortunately, in the United States, the promotion of a lower level individual does not depend upon the death or retirement of the professor as it usually does in European universities. Thus promotion of lesser quality minorities has not stopped the promotion of the more worthy although it has slowed the process.

Promotion in academia is a matter of politics. While in the government, I have had senior officials say without reservations. "You will not get a promotion or an achievement award until I do." In universities the reaction is usually not so overt but is equally damning. Committees usually make such decisions within departments, and it is not unheard of that a promotion is undercut by snide remarks or a black ball. I can think of at least one department where a competent individual was denied the chairmanship for many years on such a basis. Politics have been a part of my promotions as well. As an associate professor I was promoted to professor because I refused to become chairman until the promotion was guaranteed and there was no one else available to be chairman.

Promotion often depends upon the willingness of one to move and that, of course, depends upon a valid offer from another location. Several moves have been made in my career by academic blackmail; a job was available at a better salary or in a better location and I accepted it after dickering with my ongoing superior. At least once I have set requirements so high that I thought they were impossible to meet, only to have them accepted. On the other hand I have twice accepted a demotion in order to take a better (?) position. From Pittsburgh (Chapter 5) there was a move to the University of North Carolina at a slightly lower salary to resume an academic career. And I accepted a demotion from a supergrade position in the Federal government to remain in an academic career.

My academic career has spanned fifty years of teaching and research. I began as an instructor or laboratory assistant at the University of Texas under E.J. Lund. I then taught physics at Southwest Texas State University and radioengineering for the

Army Air force. A stint at Rutgers teaching physical chemistry was followed by teaching at the University of Pittsburgh. After two years at Mellon, I returned to full time academia at the University of North Carolina The University of North Carolina was a fine university. It was situated in a small town, Chapel Hill, dominated by the university that also owned the telephone, laundry, and electrical services. I became an Assistant Professor in the School of Medicine.

The medical school was dominated by the dean, Dr. Reese Berryhill who was a tall lanky southerner of the old school. When the antisegregation movement started, Berryhill remarked that "a Negro (there were no blacks in those days) will be admitted to this medical school over my dead body." It was almost accomplished.

I taught physiology to medical and dental students. One day Major Ted Domanski came into my laboratory. He was from the Department of Defense in Washington and he asked me about the possibility of a contract to investigate space flight by looking at the effects of long distance and long duration flights on physiological parameters of the pilots. The research was obviously leading to space flight in the distant future. One of the professors at Carolina was Augustus Miller. Gus was interested in exercise physiology and stress and we collaborated on the project. The best test for stress of the pilots was a determination of adrenal function. We did not have enough information on steroid analysis at the time to perform reasonable tests and the experiment came to an end with little to show for it except for later benefits and further Air Force contracts.

My association with Gus lead to several publications. While at NIH I conceived the idea of a textbook on endocrinology for medical students that was published by F.A. Davis and went through two editions. F.A. Davis then asked me to be editor of a series of similar books in various aspects of physiology. Gus wrote one on physiology, and the department chairman, John Ferguson, wrote on blood. In the days before molecular genetics, errors in metabolism were detected clinically and there was no way to tie them to the DNA structure. Also at that time (1950), Kenneth Brinkhouse, in the Department of Pathlogy, had a strain of dogs which had hemophilia. Although I did not work with Ken, the puzzle of how to determine the site of the inborn error of metabolism intrigued me and led to later investigations on the location of pituitary and adrenal hormones.

At the time the University was soliciting individuals who would like to spend a summer at Oak Ridge, Tennessee. Oak Ridge was the site of the production plant for isotope separation for the atomic bomb and had just been declassified in part. The officials there were interested in starting new projects in the civilian, peaceful areas and formed the Oak Ridge Institute of Nuclear Studies (ORINS) with a group of fifty universities, mostly in the South. The University of North Carolina was a member and selected representatives to spend research time at Oak Ridge. I accepted and went to the Medical Division at Oak Ridge which was directed by Marshall Brucer.

The town of Oak Ridge was almost nonexistent and we lived in barracks run by the Oak Ridge contractor. I went to work at determining the penetration of radiation into bone. We used a large caesium source and measured the penetration by determining the loss of energy through various lengths of bone. Curves could be obtained of the length of bone vs. counts per minute on a radiation counter. The worst part of the problem was obtaining bone from a butcher shop, cutting it into precisely measured lengths, and maintaining the moisture content and preservation of the specimens. It was a long, hot summers work. But we obtained information useful in treatment of bone cancer.

Along with the other visitors, I was permitted to take the course in handling radioisotopes, physics, math, and protection offered to the scientists at Oak Ridge. I have a diploma from Old Bremstrahling University of which I am very proud. (Bremstrahling is the scatter of energy from nuclear impacts).

One of the scientists working at the Medical Division was Dr. David Bruner. I did not know Dave well but we were to meet again later. Bruner had been with the AEC for many years. In 1952 he accepted a position as Chairman of Physiology at Emory University School of Medicine. I met Dave several times at Oak Ridge and we talked about medical education. He immediately offered me a position as Associate Professor in the department and I accepted. We moved to Atlanta in September of 1952.

Emory University Medical School already had a reputation as one of the better schools. It was supported in part by the CocoCola@ fortune. Dr. Arthur Richardson, the Dean would continue to become the longest serving dean of any medical school. The medical school had a very unusual arrangement. The basic sciences had its own dean or director, in this case Carl Pfieffer, who was also chairman of pharmacology. The two were often at odds. In addition, the basic departments also taught basic sciences for the dental school. In physiology the dental teachers were hopelessly out of date and aiming toward retirement.

The Department of Physiology was almost nonexistent when we arrived and Bruner rapidly hired several people. One was Peter Stewart and his wife Babette. Another was Jim Lott who proved to be unsatisfactory and I later was forced to let him go. The group was a mixed bag and we did not fool the medical students who comprise an astute group. One day there appeared on the bulletin board a piece of doggerel imitating Omar Khayyam which well characterized the faculty.

> A caravan of blind men who cannot
> Tread an arid road that they forgot
>   Cry out-"Oh, give the eyeless aspirin
> To quench the pain of that sorry LOTT
>
> Thou man called Peter, Spirit of the orthodox
> Upon whose needle dance many a paradox
>   Thy dialectic metaphysiology
> Cannot build on oh so many rocks
>
> O Brown, who knoweth the truth so well to shout it
> Then teach the believing student how to doubt it
>   Thou must two and two together put
> In such a way that one must think about it

They were accurte characterizations.

Dave brought with him a large grant from the AEC to measure the distribution of various isotopes in body organs. He set up a massive operation in which there was an assembly line production of data as a result of constant exact repetition of experimental detail.

I began to develop a research base. We discovered the drug o,p-DDD and tested it in human patients (Chapter 4) and began fractionation of pituitary glands and continued work with space physiology. The Lockheed company built the C-57, one of the largest planes in the world in a huge assembly line in a gigantic hangar in Atlanta. I was approached by the manager who suggested that man was going into space and Lockheed would like to be among the first to support the NASA projects. He wanted to build a chamber at Lockheed in which men could be isolated for days to months. The proposal was that I supervise the project, do the chemical and endocrine analysis which would be required to measure the status of the personnel, and serve as a consultant to the project. A room was constructed in the corner of the hangar. It was lost under the wings of the planes under construction. It was sealed with airlocks for transference of foods and wastes. Living accommodations were provided for two to three men. In an adjacent room we set up a small laboratory to collect and prepare samples which could then be transported the fifteen miles to Emory School of Medicine for adrenal steroid, adrenal epinephrine, and other analyses. Unfortunately the project never got off the ground. Lockheed counted on a NASA contract to perform the experiment and it was not forthcoming. The laboratory was shut down.

Conflicts began to develop in the department. Peter Stewart was a true basic scientist working on slime molds and he could not understand the research program which Bruner was maintaining. To him it was a mere testing program and the work of a technician. Bruner was interested in the distribution of isotopes in organisms determined by a strict protocol. He was testing new isotopes produced by the AEC and the analyses required meticulous care. In one sense it was testing, but in another Bruner was finding new information on the distribution of isotopes in living tissue with each experiment. Without this sort of research we would not have iodine therapy for the thyroid and a myriad of other clinical treatments. It was the distribution experiments which demonstrated that iodine accumulated in the thyroid cells and not in other cells of the body, thus permitting selective therapy. I was in the middle between the two because I was interested in the mechanism of steroid and cell activity and in the practical applications of basic research in clinical practice. The department was thus split three ways.

During the same period my attention was turned to fractionation of the pituitary gland. The adrenal responses were, of course, conditioned by ACTH produced by the pituitary. It appeared obvious that each of the various cell types of the pituitary might produce different hormones, but little was known about the site of synthesis. Separation of the individual pituitary cells might provide the answer. At Emory, the Department of Biochemistry was chaired by Alfred Wilhelmi, well known for his separation of protein hormones. The department also had a professor, Jane Russell, wife of Alfred and a famous endocrinologist in her own right. They were interested in the separation of protein hormones and it occurred to me that an interesting line of attack might be to find where the hormones were located in the cells of the pituitary and which cells produced or stored which hormones. In order to perfrom the separation, pituitary glands were obtained from the stockyards, mascerated and rapidly centrifuged in a high speed ultracentrifuge and then separated further to obtain different cell types. The separated cells were broken and centrifuged to obtain fractions which were assayed for the hormones. Wilhelmi provided valuable advice.

The fractions were identified histologically by Melvin Hess in the Department of Anatomy. A graduate student, Frank LaBella in anatomy was unable to get along with the chairman of Anatomy, Geoffery Bourne, and asked to move to my laboratory. He tested each faction with enzyme analyses and developed a profile for the location of the various pituitary hormones. The results were published in several papers in *Endocrinology.* Later, Frank moved to the University of Manitoba and received the highest award of the Pharmacological societies of the world (the Abel Award) for separation of the posterior pituitary hormones using the same technique that we developed. I won a Sigma Xi Research Award for the work. Frode Ulvedal also worked on the problem as did Carlos Nunoz from the University of Lima and Adolpho Portelo of the University of Buenos Aires who worked in my laboratory for a year.

Meanwhile, the conflicts within the department grew larger. Stewart went to Carl Pfieffer and convinced him that Bruner was not a good chairman. I defended Bruner but did so too weakly. In hindsight, I wish I had been more forceful. On the other hand, if I had, I would have probably stayed at Emory and later developments might not have been as interesting. Carl removed Bruner from the chair and appointed me as Chairman in his place. Bruner immediately resumed his old position at the AEC and moved to Washington.

When I was appointed, Carl informed me that the university would seek an outside chairman because of the departmental upset. The search began immediately. One of the first candidates and the one selected was Gerhard Brecher. Brecher was a typical Prussian. He was small, had a large bald head with a fringe of hair and pranced as he walked. He was dictatorial with students and staff. He lasted about two years.

There was a major problem with the appointment. My wife, Jessie, thought that I should have been given the position despite the reasons advanced by Pfieffer and she went after Brecher. By looking at his curriculum vitae she was able to find out that he had been in Germany during the war and could not explain what he was doing during World War II. She came from a German community which had not demonstrated whole hearted support for the war and she had served on the draft board in the community. She immediately had suspicions about the circumstances of Brecher's life. She indirectly raised questions on the campus about his loyalty to such an extent that I was called into Carl's office and asked to keep her quiet. I was placed in an awkward position with Carl, Brecher, and the school. In another situation with a common dean between the clinical and basic sciences, the situation might not have developed. Politics intervened. Art Richardson told me that he would have given me the position but he felt that he could not interfere with the basic sciences. This lead to my departure.

An interesting sideline developed while I was at Emory. I was visited by both the CIA and the World Health Organization. WHO in the entity of PAHO (Pan American Health Organization) asked me to visit and report on medical schools in Central and South America. This may have been a byproduct of the work on DDD because I had received many requests to speak at medical schools in Peru, Chile, Argentina, Guatemala and Brazil. When the PAHO invitation came, I was also visited by a CIA representative who asked me to report on any unusual activities relating to the Communists which I might see on my trip. On this and on subsequent trips, I reported to the CIA, usually on trivial incidents.

Although I was later to be a critic of American medical schools, I became convinced on this trip that we had the best medical schools in the world. I was appalled to go into the medical schools in Honduras and see the medical library as a pile of books in the corner of a room. I was horrified by information from students that anyone could graduate or enter medical school as a democratic right and could remain in school for any number of years in order to finish. I was equally appalled in Buenos Aires which had good people (Houssay, the Nobel Prize winner in physiology, for example) but thousands of medical students because no screening was done to eliminate the incompetent. In Santiago, Chile the Universidad Catolica had a medical school but almost no equipment and Croxatto in Physiology made do with makeshift material. In Lima, Peru, one of the world's great respiratory physiologists, Hurtado, had offices in what looked like a down town commercial building. The medical schools were unimpressive compared to the Far East and Europe.

A major extracurricular event occurred in Honduras. During my stay , I was awakened by a popping beneath my window. I looked out to see a soldier firing an automatic rifle up the street. A revolution was in full swing. There were only two buildings over two stories in height in Tegucigalpa at the time and one was the hotel in which I was staying. The other was a rival hotel several blocks away. The government placed heavy guns on one roof top while the rebels placed guns on the other and they fired at each other. Fortunately, I was in the government controlled hotel-they had bigger guns and better trained soldiers. They rapidly subdued the rebels although we were forced to stay in hiding in the interior of the hotel for a day or so. There was an interesting connection with my role with PAHO. I met the President of Honduras who was a pediatrician to be both reassured about the revolution and to talk about medical education. He was the first physician I had seen with an automatic rifle standing in the corner of his office and a pistol on his desk

The fiasco with Pfieffer lead to my departure. I met a former NIH Director, Eugene Dyer, and he interested me in moving to Washington (Chapter 4) where I taught at both Georgetown and George Washington Universities Medical Schools. My most interesting acquaintance was Estell Ramey, professor of physiology at Georgetown. Estell was an endocrinologist with a beautiful face and heavy legs. She was a female activist and later obtained a reputation as an outspoken advocate of Women's rights. She and I split the endocrine lectures. When I left the government (Chapter 2) I returned to the university, first in San Antonio as the head of the Southwest Research Consortium (Chapter 8). I held three teaching positions. As Professor of Physiology, I taught in the Medical Physiology course and occasionally taught a course in physiology for nurses. As professor of Environmental Engineering at the University of Texas, I taught courses on the environment, and as Adjunct Professor of Public Health Administration at Trinity University, I lectured on occassion.

In 1978, I was offered the position of Associate Provost for Research at the University of Houston. With it came an appointment as Professor of Biology. Early on, contacts were made with the Medical Center and I was rapidly offered the position of Adjunct Professor of Public Administration at the University of Texas and later (Chapter 2) at Baylor college of Medicine. The Texas Woman's University had a large branch at Houston and offered a Master's degree in Hospital Administration. I was offered the position of Professor of Hospital Administration and for three or four years taught regularly in the graduate school. Then there was a major shake up in the administration. The Director at Houston became ill and I was asked to come to Denton and be interviewed for the position. Texas Woman's University was located in Denton Texas, several hundred miles away and the governance of the Houston campus was spotty at best. I may have been too outspoken about the problems and the need for major change for I was not offered the job.

In the meantime the President of the University of Houston, Barry Munitz and his deputy, Andrew Rudnick, parted with me at their request. Attempting to control the Deans (Chapter 13) who wielded too much power was impossible. The Provost George Magner suggested a return to a full time teaching position in the Department of Biology. This was no difficulty as I had been teaching in the department every semester since coming to Houston.

A research program was maintained at Houston during my tenure. Although the major effort was the use of the laser card in patient record systems (Chapter 2) I had a grant from NASA, a grant from the EPA to explore energy sources, and small university grants for minor projects. One student finished a Master's degree working on computer development and a second obtained the Ph.D. degree with a dissertation on artificial hearing devices.

In 1990, retirement came at seventy years of age along with immediate selection as Professor Emeritus. Heavy involvement in research still goes on, continuing more than fifty years in teaching and research. I am proud of my long career and where it has led. My life in academic and industrial research and in the government provided a priceless experience not to be exchanged for any other.

CHAPTER 13

POLITICS AND PROMOTION

Politics in science occurs in both academia and government. A new administrator makes haste to clear out the predecessors people and replace them with his/her own as rapidly as possible. This is particularly prevalent in Washington where several thousand higher administrative officials are replaced at each change of administration. Although the higher Civil Service grades overtly are not disturbed, it often becomes untenable for one to remain in the face of opposition.

Emory Medical School in Atlanta was unique. It was organized into two medical schools, one in the Basic Sciences headed by Carl Pfieffer and one in the Clinical Sciences headed by Art Richardson. They were often in conflict for money, positions, and space. Fortunately, Richardson had the ear of Boiseflieut Jones, from the Woodruff Foundation that was supported by the total CocaCola® resources and was able to dominate discussions. The Department of Physiology at Emory was required to teach both medical and dental students. We have mentioned the politics of that situation in Chapter 12 Divided responsibility cannot succeed.

There occurred one of those fortuitous events that so often happen. Emory had asked a former director of NIH, Eugene Dyer, to take a position as a general advisor to the medical center. He was a little man with white sparse hair, enormous thick glasses to correct for cataract removal, and a wide knowledge of health affairs. Gene and I became friends and we played bridge frequently. At a bridge game one night, he had a visitor who did not play. He was introduced to me as Fred Stone, Deputy Director of the Division of General Medical Sciences at NIH. Fred came by the office and talked to me at length. In the end he offered me the position of Executive Secretary, Physiology Training Committee at NIH. In those days before the dramatic rise in university salaries the government paid more than the university and the job looked interesting. I accepted.

NIH was a revelation. The Institutes were set in a lovely landscaped, tree swept campus in Bethesda several miles outside Washington. There were not as many buildings and they were less impressive than the present campus. But it was the time of Lister Hill who handled health appropriations in the Senate and John Fogarty who handled them in the House. Both were astute politicians and dedicated to expansion of health services. In addition, Mary Lasker who had a vast fortune made from advertising, worked diligently to advance health causes. Each worked to obtain more money for health research. Politics really worked.

One of my first jobs was to try to increase the research capability of the black schools of the country. The original idea was a purely political one to gain support for the expansion of training programs in the basic and clinical sciences. We recruited from California Dr. Jerry Woods, a well known black researcher and friend of Martin Luther King, and she and I visited a large number of black schools, offered grants to the faculty to go away for further training, offered students fellowships to continue in school with good grades and offered to fund research equipment We found a dismal situation. Ill equipped labs, professors equally ill equipped for modern research and with little desire to improve were common. Although this was the first program at NIH, and I believe, in the Federal Government, it was not successful except in so far as we expressed an interest in black students and schools.

During this period NIH was expanding far beyond its ability to provide space. In ten years I moved five times. The Division (DGMS) was housed in the home of the Bishop who had given the land for NIH to the government, the Stone House. It was a fine old mansion with high ceilings, corniced rooms, and fine woodwork. The Director, Halsey Hunt, had his office in the library endowed with a great stone fireplace and dark woods. The bathrooms contained filing cabinets and the closets were filled with old records. Within a short time we moved to the Trunnel building in downtown Bethesda over a laundry and dry cleaning establishment, then out of Bethesda to a small complex of buildings, back to the Westwood building between NIH and the city of Washington, a huge government building holding much of the functions of grant administration for NIH, and finally to the administrative building on the NIH campus (Building #31). The moving problem was political. It represented the efforts of the government to lease space at the lowest possible costs regardless of real needs.

It was an exciting time. We were starting biomedical engineering (see Chapter 3) and developing new resources. Fred Stone was a natural administrator who had the ability to foresee directions of sciences and lead men to adopt them. He was short, very heavy set, decisive. He had been a marine during World War II and it had never worn off. His trousers were always sharply creased, he wore brilliantly shined boots and his suits were always cut the same old fashioned way. He backed those who produced; because of his push and support I went from Executive Secretary to Acting Director of an Institute in less than 10 years.

We accumulated a group of highly sensitive and competent individuals, most of whom developed great reputations. I remember Robert Livingston who came to be head of a branch and then moved on as chairman of the Department of Neurology at University of California San Diego, James F. Dickson whose name is synonymous with biomedical engineering, Helen Schroeder who became a senior official in NHI headquarters, and others.

In one sense, Fred Stone and I served as trouble shooters. While in the DGMS we had organized biomedical engineering training, started training in clinical sciences, and developed new areas in the basic sciences (Genetics). Then Stone and I were asked to move to the Division of Research Resources. The Division had a variety of activities with the general purpose of providing any necessary adjunct to research. There was a branch for research construction, for animals used in research, for technical support, for support of regional medical programs, and for clinical research centers. All had problems. The construction branch that built research laboratories was lead by Frank Schmell. Frank had been with the group too long and was more closely tied to the universities and hospitals than to the NIH. He was willing to approve any expenditure that the university thought proper. In one, but much smaller, sense his activity resembled that of the universities now under fire for expending Federal funds for unnecessary costs. We had to curtail marble walls, fountains, and decorative features to the consternation of Schmell and the universities. I served as an interim administrator in the branch.

The Clinical Centers Branch was managed by Sam Silberman, a psychiatrist, who became unable to continue to make decisions. I became acting director at a time when the General Accounting Office found that the clinical centers, designed for research in hospitals, were treating patients who were not on research protocols. I was told to go out and recover the funds. The resistance was monumental. Some hospitals or medical schools would have been required to refund in the millions of dollars. Politics entered the fray. The larger and more prestigious institutions went to their congressman with requests that bills be passed to forbid recovery of funds. No funds were collected.

The Animal Resources Branch was also under scrutiny. Hal Eyestone, the director, was a veterinarian who was charged with the responsibility for establishing several primate centers to do research on human diseases using primates. The centers involved a major construction of facilities, employment of a large staff, research support, and related activity. The major component of each project involved the appointment of a director who could oversee the entire project. There were problems, Harvard insisted on appointing a director who was not satisfactory to NIH and other directors were replaced during the project. It was difficult to secure funds from other institutes for research in the centers. We had many application for centers and they had to be site visited, and ranked. It was a long and painful process because each university thought it was the ideal place for a center. Bill Gay took over as a director of the center after I had served as interim director and the primate centers were set up and are still in operation.

The Special Research Resources Branch was less of a problem. I served as the branch chief at the same time acting as Associate Director of DRR (Division of Research Resources). We started the computer project at MIT (Chapter 2), and a variety of other special resources such as the Horning Mass Spectroscopy Center at Baylor College of Medicine. One of the most interesting projects was at Oak Ridge Institute, a division of the AEC. There I met Al Weinberg, the director, and we arranged to fund Norman Anderson, a quiet, sober scientist interested in developing special analytical instruments. Norm developed a means of analysis of blood samples similar to the process used in most clinical laboratories but occupying one tenth of the space. It used a unique process of mixing the samples with the chemicals by centrifugation in a very small centrifuge. It was placed on NASA flights to meet space and weight requirements. Norm went on to win numerous national awards.

In 1968, Fred and I were asked to move back to General Medical Sciences, now an Institute. As Associate Director I had responsibility for all operations as there was no other Associate or Deputy Director. We continued to develop research, expanding into clinical research as well as clinical training. Dr. George Jordan, Professor of Surgery at Baylor College of Medicine, came to spend a year developing research in various aspects of surgery. Not incidentally, George's boss was Mike DeBakey, Professor of Surgery at Baylor, who had headed the President's Commission on Heart Disease, Cancer, and Stroke and had the direct ear of the White House.

I was asked to testify before Congress several times on the NIGMS budget. It was here that I was at cross purposes with the NIH Director, Robert Marston. Daniel Flood, chairman of the committee that handled NIH appropriations had been briefed to ask me about development in biomedical engineering. I just happened to have in my briefcase a series of biological amplifiers made by Wen Ko of Case Western Reserve University all of which had the same function but ranged in size from large vacuum tubes to a microchip. Flood asked me to explain them spite of Marston's objections and I answered. The cross fire created problems later.

The period of the 60's was a great one for NIH. Budgets expanded each year in staggering amounts. There was a set formula. When one appeared before a Senate or House committee, the budget proposed by the Secretary or the President was laid out and no one was permitted to deviate from it. However, there was a catch. If Flood or Hill looked over the table and asked a direct question: "Now doctor, what do you really need to do this job", we were permitted to answer. In that way budgets were raised each year. In addition, Mary Lasker knew a large number of congressmen and Senators and had contributed heavily to their campaigns. She maintained a house in Georgetown in addition to her regular establishment on Beekman Place in New York City. It was not unusual for her to receive a copy of both the proposed budget and the budget which NIH would like to see on the day before the hearing. I have been to her house in Georgetown many times carrying the proposed budget and a list of projects we would like to see funded She would make the necessary visits or phone calls to speed the process along. Mary would often come over to NIH and we would sit in the cafeteria and over lunch decide on a strategy of which projects were most likely to get funded. On rare occasions, usually at their instigation, we would meet in secrecy with members of Congress or their immediate staffs. Harley Dirkson, the administrative assistant on the committee handling health appropriations knew more about the NIH budget and more about the general developments in health across the country than NIH staffers. He was a pleasure to talk to because he was willing to say that this project could be supported while another would not. This procedure was disapproved of by the powers that be and could have severe consequences if revealed. But it was helpful to know what the member thought of the budget, which items he would be most likely to support, and to have an opportunity to pitch for a particular item.

There were other inputs to the budget process. Each Institute had an Advisory Council and there was a National Advisory Health Council. The members were originally chosen for political acumen, scientific ability, and knowledge of the system. In the beginning most individuals were nominated for scientific ability and no consideration was made of party or political affiliations. This changed in the

1970's. Each nominee had to be submitted to the White House which could reject or propose other nominees. The degree to which this was observed depended upon the administration. Nixon scrutinized every person and often rejected one. He insisted on knowing the political affiliation of every nominee and reserved the privilege of rejecting those of opposing views. Kennedy was much more lenient. Johnson was much like Nixon. But every effort was made by the Institutes to nominate individuals who would understand and support the program to the fullest. I can remember meetings of the Advisory Council with Phil Handler, Mary Lasker, Mike Debakey and other well known personalities present. The National Cancer Institute (NCI) had Danny Thomas on its panel and NIGMS had Jerry Woods a prominent leader of the civil rights movement and a friend of Martin Luther KIng.

Yet decisions were usually well taken. However, there was the inevitable parochialism. A member would defend a friend or an institution. Defenses were often made by the lay members on political grounds. Although any member representing an institution with a grant before the panel had to leave the room, feelings could be obliquely expressed. The council had the right to reject any application but not to approve one which had been previously disapproved without further investigation. Happily rereviews were not often suggested.

Some members of Congress were noted for their aggressive approach to funding. We have remarked on the role of Spiro Agnew in funding EMS Systems(Chapter 6). When we established the STARPAHC project with NASA (Chapter 7) Senator Jennings called me into his office and demanded that we fund a train which would go through the mountains of West Virginia, equipped with instrumentation to perform the same tasks the vans performed in the NASA/DHHS project. We tried to point out the costs and the difficulties. A famous Senator from the West Coast was interested in developing health centers in his region. He had no compunction about calling to say, "I'll put $50 million in your bill for ---centers provided that $10 million goes to ---." He sometimes got his way.

One of the best politicians I met in Washington was Annie Wanetka, on the tribal council of the Navajo tribe. She appeared in HSMHA (we administered the Indian Health Service) a big, tall woman, dressed in black with skirts dragging the floor, wearing pounds of spectacular turquoise jewelry to make appeals for better health care. She was a skilled speaker, a consummate politician, and was usually successful.

HSMHA was divided into the Health Services Administration and the Health Research Administration and the functions were distributed between the two areas in 1973. Kenneth Endicott, formerly head of the National Cancer Institute was selected as head of the HRA where I would be stationed. Endicott and I had known each other at NIH and I had little respect for him. He was a boorish, a heavy drinker, uncouth in manner, foul mouthed in speech and not a very good administrator. He had obtained positions by political connections on the Hill. It was said that his deputy, Carl Baker ran the Institute. It was apparent that I could not stay and the departure was hastened by the appointment of Ken's own people in key positions as might have been expected.

Harold Vogtborg, the President of the Southwest Research Institute and the

Southwest Foundation in San Antonio, Texas had repeatedly suggested that I should consider a move to San Antonio. Vogtborg was a competent engineer and the Southwest Research Institute was one of the largest nonacademic research institutions in the country. He wanted to pull together the research interests of San Antonio, and although he died very soon, the Southwest Research Consortium was organized with Frank Harrison, President of the University of Texas Medical Center, Peter Flawn, President of the University of Texas and Martin Goland, the new president of the Southwest Research Institute as the Board of Directors. I was asked to come as coordinator and organizer. The political weight of the group in Washington produced a situation in which I would be sent to San Antonio for two years to organize the consortium and my salary would be split between the San Antonio group and the Federal Government. I moved to San Antonio in the Fall of 1974. My future and that of the consortium was unclear after the assured two years. Actually, the contract was renewed for an additional two years.

I spent large amounts of time in trying to identify areas of research between the institutions in which cooperation was possible. Success was moderate. I helped Henry McGill in the Department of Pathology at the medical school prepare a grant to the NCI for the study of the causes of cancer as a result of smoking. The Southwest Foundation had a large primate colony of mostly baboons and they were taught to smoke cigarettes for a reward. The grant running into millions was finally paid. McGIll later moved to the Foundation to become Director of Research. We also tried to set up a cooperative research project between Carl Pauerstein in Ob/Gyn to study ovulation at the Institute but it was not as great a success. The problem was partly Pauerstein's ego. And it was not his alone. Many physician-scientists had the "god" complex. They were unwilling to give anyone credit for anything they could do themselves even if the results were not as good. As a result, each wanted his own research grant and all of the credit and hesitated to join a consortium which might have produced more research but less power.

The Federal government had become concerned about medical practice in which the physician failed to inform the patient about risks of treatment and decided that informed consent was necessary for any procedure on a patient. The physician was required to explain to the patient every detail of what would happen to him/her and the patient then signed a paper certifying that the medical procedures were fully understood. The consent forms were honored more in the breach than in the reality.

The consent forms were reviewed by a Review Board for content, and complaints against the system or the physicians by the patients were referred to the board for action. Records were not organized, meetings were sporadic, and a definite policy for review had not been developed. I organized two review boards, to handle the large volume of work between the medical school and the hospitals, and tried to organize the board for greater efficiency. Other hospitals, the University, and colleges in town applied for membership and we organized a review system for the area around San Antonio. We devised a more efficient system for review of applications for research, changed the consent forms, and streamlined the procedures. The results were published in a series of papers on the mechanics, costs, and policy of review boards which attracted wide attention.

By the time my term of detail duty ran out in the Federal government, I had twenty five years of service and could retire. I choose to do so. It was also becoming apparent to the consortium Board and to me that we were unlikely to achieve a real consortium of several institutions aimed at a common goal. Parochialism was the basic reason.

The University of Houston approached me at the time with an offer to become the Associate Provost for Research and Development. We moved to Houston in 1978. The President was Barry Munitz who had assumed the office from Philip Hoffman. Hoffman had grandiose ideas and had attempted to build the university from a large urban campus to a system by adding a downtown campus, a campus at Clear Lake a few miles from Houston and a campus at Victoria on the Gulf coast. We were now trying to find the accommodations which would make such a diverse system work. The Provost was George Magner. George was a tall elegant man with white beard and hair. He loved martinis, the job of Provost, and social work in that order. He administered everything and faced every crisis in the same urbane, low key fashion.

Although our main job was the academic health of the university including research and fellowship awards we were soon embroiled in a major dispute. The university had built a solar power station in the California desert to produce power by means of mirrors reflecting light and therefore heat onto a receiving surface thus producing steam and power. The funding for the research project was through the EPA. As the project developed, Georgia Institute of Technology in Atlanta became interested in a similar project. EPA informed us that we would have to cooperate if either was to receive funding. I spent many days on planes between Atlanta and Houston and occasionally Washington trying to work out a compromise which would allow both parties to obtain credit and funds. Each project director tried to outdo the other in terms of the amount of funding by going over the university heads to the Federal government. The project managed to get off the ground in spite of the controversy.

Conflict soon arose with the deans of the colleges. They had been almost autonomous under previous administrations but it was necessary to gain some control over them. The College of Engineering had a large number of graduate students who were also used as teaching assistants. There was a constant stream of complaints that the graduate students were 85 percent foreign, many spoke little or no English, and most had no previous teaching experience. Students were learning little in the laboratories and review sessions. We attempted to impose a rule that any teaching assistant should be able to communicate with his students. Naturally other deans objected to becoming a training ground for teaching English while the engineering dean objected to rejection of graduate students who were supported as assistants. We then found that many graduate students were supported on several grants, each reporting full time effort for the student. The GRE (Graduate Record Examination) was used as a standard of admission to colleges but was neglected when it came time to make appointments.

The Provost's office had two committees, the Undergraduate Council and the Graduate Council which were supposed to oversee these matters and make policy which could then be adopted or rejected by the Provost. The committees were composed of a member from each college, elected by the faculty and this produced an automatic political climate. Members from the "have-not" colleges (fine arts, humanities) opposed the members from the "have" colleges (the sciences) on standards for graduate students, use of teaching assistants, and similar matters. It was a no-win situation.

This was a time of great political and sociological unrest. As a large university we maintained several animal colonies and conducted work requiring animals. The School of Optometry used monkeys to test various eye preparations, engineering was testing a new heart valve design, and others were exploring various parameters of physiological response. Aroused students began a campaign to halt animal research. The university took the position that such work was necessary, that the experiments were reviewed by national committees, and that animal care met all Federal requirements and therefore the work would go on. The campus newspaper continued to be filled with diatribes from unknowing students. I had been appointed to the Animal Care Committee of the American Physiological Society and used this platform in reply to the antivivisectionists in the *Cougar*, the campus newspaper. The problem is not yet resolved.

Appointments to the faculty were also under scrutiny. The university was under severe political pressure to recruit women and minorities. Success was not great for several reasons. There were a large number of women in the humanities, fine arts and social sciences. but there were few in the hard sciences. There were almost no blacks although there were several minority groups in the form of Chinese, Japanese, and Taiwanese in the sciences. The reason was obvious. When small numbers of blacks graduate from high school, and still smaller numbers go to college, the number continuing into graduate school is very small, and of those with a graduate degree in the sciences are still fewer. I have never been willing to relax standards to hire or promote any person who does not meet the requirements-blacks included. This attitude markedly limits the number of successful applicants for any position but if we are to compete with the world we must compete first in educa87tion before we can compete in technology.

After two years I resigned my administrative position and returned to the department of Biology as professor. The department was a hot bed of politics. At least three members thought they should be Chairman and would do a better job than the current mild mannered, quiet and sane chairman, Joe Cowles. Snide remarks, backbiting, and soliciting support went on in the halls of the department. As the chairman was selected by the faculty, each election resulted in politics at fever pitch. Fortunately, the number of candidates resulted in such division within the department that Cowles was reelected time after time. When Cowles left for another position, the division was so great that the Dean opted to select the chairman upon the recommendation of the faculty. Each member of the faculty paraded into the office of the dean, presented his/her views, nominated a candidate, and answered a few questions about the qualifications of the candidates as they saw them. The dean then picked his candidate who was not the favorite of the majority of the faculty.

I have always enjoyed politics as a sideline to science although in some cases the politics of the situation have gone against me. However, if science is to advance, if funding is to continue, and if new areas are to be explored, we must be politicians as well as scientists. Instances can be cited in which politics has been of great advantage to science and where it has been a major disadvantage. The recent funding of the supercollider in Texas is an example. The politics of economics and the billion dollar expenditures secured funding. On the other hand, the space station despite political pressure on one hand has been downgraded largely because of the expense. This creates a major dilemma. One the one hand big science is now costing in the billions of dollars per project and funds are hard to come in a time of straitened budgets while the little science does not attract the attention created by big science and so also does not get funded.

REFERENCES

Brown, J.H.U., L. Schonfield and P. Allen *Functions of An Institutional Review Board.* Federation Proceedings 38: 2049 (1979)

CHAPTER 14

EDUCATION

As an educator for fifty years teaching at all levels from high school to post graduate courses, seeing segregation outlawed, massive changes in curriculum and the advent of computer technology in teaching I suggest that several conclusions can be drawn from my experience.

America's shame is its schools. We have no commitment to learning, no evidence that a student is taught to his potential, and every indication that the system is failing day by day. We are in competition with the Japanese and the Europeans who consider education to be essential and insist upon quality and quantity. Our Scholastic Aptitude Test (SAT) scores trend generally downward and some 20 percent of all Americans cannot function in a literate society and probably twice that number cannot function in a technological society. Dropout rates average about 30 percent for whites, 40 percent for Latin-Americans, 50 percent for blacks, and perhaps 10 percent for Asian students. Surveys conducted by the school systems and by the Gallup poll for the National Geographic Society suggest that a large percent of Americans are illiterate when asked about geography, politics, and current affairs. American students were unable to identify 50 percent of all landmarks as compared to Swedish students who identified 90 percent. Despite the fact we have fought a war and stationed troops in the Persian Gulf, 75 percent of all American students could not find the gulf on a map. A similar percentage did not know where Vietnam was located. About one-half of the students were not able to find Japan or France on a map. Of greater concern is the fact that younger students do worse than older students on these questions, indicating that our system is deteriorating rather than improving. School in the United States rarely exceeds 5 days a week with many holidays—180 days and 36 weeks per year and the average student does about 30 minutes of homework per day. The typical Japanese student goes to school five and one-half days per week, 240 days per year in 44 weeks and does about three hours of homework per day. In addition, any Japanese student with ambition toward college usually spends an additional two to four hours per day in study, often with a special tutor.

The education accorded to youngsters does not correlate well with the money spent to develop the school systems. Texas does not rank high on the academic totem pole and it spends about $4,000 per pupil per year in grades K-12. One would assume that $4,000 would buy a good education. However, it turns out the highest ranking school scholastically, as determined by the standardized tests, receives pupil ($4,800), which ranks twelfth. Furthermore the top scholastic ranked school. has an average SAT score of 950 compared to the school spending the most money per pupil, with a SAT of about 900. Money does not necessarily secure a good education. It is certainly true that if the amount of money spent per pupil decreases below a maintenance level, then scores do correlate with available funds.

Money, the financial magazine, in May 1990 looked at two schools in Illinois and New Jersey with the same enrollment, dollars spent per pupil, resources, and family income. They found that one school sent a large proportion of its students to college and had high SAT scores while the other had many fewer college attendees low scores, and other problems. Obviously family income and expenditure per pupil were not the deciding factors. In another study reported in the February, 1992 *Scientific American* low income Asians were compared to low income blacks and whites in our schools. The Asians were the highest ranked in the schools, made the better grades and contrary to popular opinion, the members of the largest families made the best grades as did those who worked while going to school. In addition, the Asians studied more than twice as long as other groups. We must examine this type of data to find out what is wrong with our schools.

Teaching was at one time considered to be an honorable profession. Today, in the United States, it is not. Paper work, unruly pupils, and low salaries have driven the teachers from the classroom. The trends in education in the large universities have clearly demonstrated the problem.

The brightest students in science used to go into engineering and medicine. However, the brightest are no longer attracted to science and go into business. The next level of science students enter graduate school while the less competent enter dentistry, physical therapy, and similar professions. Finally, those who cannot qualify for special training go into teaching. There are always exceptions and these are eagerly looked for by faculty, but when a student arrives for counseling in science with a C average, he/she can be expected to be directed toward teaching.

This attitude is shameful in a time when teachers are in great demand. But it is largely the fault of the teachers themselves. Present data indicates 10 percent of teachers cannot add a column of numbers, 35 percent cannot convert metric system values into or from the English system, and 74 percent cannot calculate values in an algebraic equation. We need the brightest minds in the classroom. It is doubly unfortunate at the high school level that other considerations are placed above teaching, as we will discuss later. And in departments of education, emphasis is placed upon the training to be a teacher, not upon the subject to be taught.

The high school teacher should be well paid and relieved of the burden of administrative duties and after-hours trivia. We need a professional teaching system where the paperwork is delegated to assistants and the teacher is given time to learn about the modern trends in society and to advance in education.

A. Fraley has written an interesting book entitled Schooling and Innovation dealing with attempts of the public schools to solve the problems of education of the young. Unfortunately the book is largely an excuse of the school system with very little attempt to provide constructive criticism. Fraley points out, for example, that the schools are deficient in that they fail to teach the three Rs, they do not meet the needs of children and society, they do not provide equal opportunity, they do not teach democratic principles, and they do not create a better social order. In each case the blame is not placed on the schools but on the society in which we live. The point is not made that a good educational system beginning early enough might solve some of the problems. Fraley does point out that conflict between curricula as proposed by committees, school boards, and teachers are often at the root of the problem, and she implies time and money would solve most of the problems. Fraley confuses the issues as do most educators. True, we should have a better society, and it must begin in the schools, but without a fundamental background in the three Rs, rhetoric about a better society must fall on deaf student ears. H.J. Aarons, the well-known economist and philosopher, has pointed out that money is not the major problem of the school system. Real expenditure on schools has risen about 4 percent per year for twenty-five years and yet the quality of education has decreased. During the same period of time student /teacher ratios decreased from twenty-five to one to eighteen to one, more than 50 percent of teachers obtained graduate degrees, and 50 percent had 15 years or more of teaching experience. Yet the SAT scores declined from 960 to about 905 during this period. One of the problems Aarons pointed out is the bureaucracy. During the last twenty-five years, the nonteaching employees of school systems increased by 400 percent and are now more than 50 percent of the total employees

In any area of life, business, work in a fastfood restaurant, or as an engineer or scientist, mathematics is required. The ability to make change, to calculate percent, or to figure interest payments demands facility with numbers. Mathematics is useful on a personal basis, in the civics and political arena, as a professional requirement, for leisure and to develop cultural pursuits. It must be a part of every student's life. Unfortunately, our students are not trained from grade school on to handle figures.

A recent report by the National Research Council, *Everybody Counts*, has highlighted the problems. The council points out that the difficulty is not the lack of an innate mathematical ability in some students, but that students do not apply themselves, parents deprecate math as a subject, or teachers are incompetent. The report goes on to suggest that the situation can be turned for the better by a reasonable approach. However, the problems are compounded by a lack of knowledge about how to remedy the deficiency. All attempts by the educational community to date have been failures.

Further problems arise when the demographics are considered. At present, about 25 percent of whites and 50 percent of blacks fail to finish high school and therefore are deficient in math. Yet the latter group with other minorities will comprise most of the additions to the work force by 1999. Only 15 percent will be white males. The demographics are also against the teacher. Statistics indicate that there will a massive shortage of mathematics teachers with no chance of remedy.

The Boston Latin School is 350 years old. It was established in 1635 to provide for the education of future leaders of the country. Boston Latin School fulfilled the need and graduated such notables as Ralph Waldo Emerson, George Santayana, Theodore H. White, Leonard Berstein, and John Adams. Admission was by ex amination and the curriculum was so tough that attrition averaged about 30 percent. But that may be changing. The NAACP has sued to have the school opened to all so it would become just another public high school in Boston. Nothing could be worse for education in the United States and nothing could more clearly demonstrate reverse discrimination. Boston Latin School for years has stood for quality education and has made the student toe the line or leave. To argue now that blacks have a right to attend the school without qualifications makes a farce of education in the United States at a time when all forces suggest more, not less education should be our goal.

Boston Latin School standards should reflect our own. We should be proud of a certain degree of elitism. No person and no nation can advance without it. Few would argue there would be little point in presenting a hack violinist on the same stage with Itzak Perlman, but these same persons would argue that a college or a high school should treat everyone in the same way. Elitism is as important in education as it is in arts or music where it has been accepted as a norm.

The elitist attitude is accepted in Japan and most countries of Europe where a series of tough examinations weed out the poorer students in grammar school and later in high school so that only those at the very highest levels of performance can expect to enter college. In stark contrast, our educational system has been geared to a least common denominator. We spend large sums of school funds on the "special" child and neglect the really special child, the gifted pupil. It has been said with some truth that the majority of inventions and new ideas come from 5 percent of the population. It is these to whom we should pay attention. The Japanese and the European have long known this and the strict examination schedules they use from grade school to decide admission to college is an indication of their intentions.

Education is the backbone of our society. It produces an educated electorate, a competent work force, and innovations to develop the future. Without it we are a third-rate nation. With it we can again compete for first class. Time is running out. A whole generation of educationless students has been created, and another generation is beginning. If we do not act now to upgrade teachers, curriculum, and schools it will soon be too late to save still another cohort of students.

It is imperative that industry demand standards of the school system to produce a quality workman. Pressure can be applied through the force of taxation. Industry as a whole pays a large portion of taxes and thus should have a say. Industry can help in other ways. Experts in various fields can be loaned to the school system, magnet schools can be encouraged in areas of interest to business, funds can be provided to develop new curricula, and books and instruments can be donated to laboratories. For example, organization of consortia for development of educational ideas between academia and industry could be encouraged.

REFERENCES

National Geographic Society, *Knowledge of Geography*. (Preprint) Washington, D.C: National Geographic Society Press, 1989.

National Research Council, *Everybody Counts.* Washington, D.C.: National

Academy of Sciences Press, 1989.

Brown, J.H.U., and J. Comola. *Educating for Excellence*. Greenwood Publ. Group. Westport, CT, 1991.

CHAPTER 15

ACADEMIC HONORS

Every scientist wants something besides money, publications, and promotions. Everyone wants to have recognition for their efforts as determined by peers. Promotion often occurs because of tenure and with no recognition of merit. Salary raises are sometimes determined by legislatures. Publications may not be recognized. But the award of honors for research calls attention to the effort, recognizes the scientist, and calls the achievement to the attention to his fellow workers.

Rewards may be of several kinds. Election to an honor society ranks a person with peers. Selection to an organization based on merit is another measure of success. Selection by a university to a Distinguished Professorship or similar title confers a local fame. Elevation to the fellow rank in a major scientific society clearly marks the individual as above the average. Of still greater importance is a national or international award. These rank from election to the National Academy of Sciences, the Lasker award, and the National Medal of Science, to the Nobel Prize. Awards may also be given by a local organization or a national society but there is a difference in the way they are recognized by peers.

Selection of an individual to an award takes place through many pathways. The normal way is by nomination followed by review of a committee. The route varies. The National Academies require that nominations come only from fellow academicians. There is supposed to be strict confidentiality. However, I have received letters from individuals suggesting that I agree to nominate or second their nomination. It is not too unusual for individuals to agree to promote a nomination with the agreement that when the first is elected, he/she will then propose the second for membership. On the other hand, the Nobel Prize committee solicits nominations from a range of individuals world wide. I have helped to prepare recommendations or nominations for renowned scientists at NIH.

The numbers of members of any organization may be limited which restricts the nominations. The IEEE has a certain percentage of Fellows and this number is relatively constant. The national academies establish a rate of growth and attempt to stick to it. The Cosmos Club, the major club in the United States recognizing excellence in its member, has a membership limit by bylaw and new members may be admitted only when a vacancy exists. The national honor societies have a standard, usually grade point average, number of performances, or other criteria and anyone who meets the standard is automatically elected. Some organizations admit anyone and restrict selection only to those who are awarded the Fellow grade. The American Association for the Advancement of Science is a good example. Anyone can join, subscription to the journal is the only real requirement, but some distinction is required for admission to the grade of Fellow.

Nomination for academic honors generally proceeds through a time frame. The individual is elected to honor societies, he is recognized by his college with promoion, he may become a fellow in a scientific society, win a medal or citation, and with luck and ability he may be elected to one of the more prestigious organizations where mere selection is acknowledged to recognize quality.

Numbers may be an indication of the importance of the honor. There are tens of thousands of members of Phi Beta Kappa, there are hundreds to thousands of members elected as Fellows in the scientific societies, and there are hundreds elected to membership in the national academies.

Selection to a national honor is not always a positive choice. Scientists who do not wish to be outmoded or who wish to retard progress of another have been known to deliberately black ball a nominee. I once selected for nomination to the National Academy a professor recognized by his peers who were willing to support him fully. The dean at his university who had hired him in the first place and then made him chairman of department refused to support the nomination on the basis that "I do not know him well enough." Strangely the dean was also a member of the National Academy but in a different field and he was trying at the same time to obtain election for a colleague in his field.

Conflicts concerning selection of candidates may arise from strange sources. In those organizations where there is a strong industrial component as well as an academic one, selection may be based on concepts of "basic" versus '"applied" science. The National Academy of Engineering has faced this dilemma at every meeting. The National Academy of Sciences is now facing it with nomination of individuals from genetic engineering companies who are competent scientists but have the taint of applied science. But times are changing.

The honor of belonging to a society may change with time. At one time the New York Academy of Science carefully selected members and chose only those meeting stiff criteria. Now the Academy solicits membership from anyone. The National Geographic Society advertises memberships, implying high standards but actually anyone who subscribes to the *National Geographic Magazin*e is a member.

A form of honor which is being exaggerated is the inclusion of a biography in "Who's Who" Although there are standards for inclusion, the standards are not high. For example, at least one publisher accepts all professors regardless of demonstrated ability. Some publishers attempt to tie inclusion in the Who's Who to purchase of a copy of the volume or a plaque celebrating the notice. We are now seeing the birth of dozens of Who's Who in sciences, engineering, the arts, nuclear energy, general information and no one knows what else as the desire to be recognized increases.

I serve as an illustration of most of the discussion above. While in college I was elected to Alpha Chi, the national honor society, Phi Kappa Delta, the national forensic honor society, to Purple Mask, a national honor theatrical society, Phi Lambda Upsilon for national honors in chemistry, the American Chemical Society, and Sigma Xi, the national scientific society. In graduate school I added the New York Academy of Science, the Gerard Swope Award, and the Society for Experimental Biology and Medicine all selected on the basis of demonstrated publications. As a professor I became a member of the Endocrine Society, the American Physiological Society, and the IEEE. After a few more publications, I became a Senior member of the American Chemical Society (the grade no longer exists), and a Fellow of the AAAS. With the publication of several books I became a Fellow of the IEEE. During government service I received a Special Award from NASA, and a Certificate of Excellence from NIH. My scientific political career was established with the organization and presidency of the Biomedical Engineering Society and the formation and presidency of the Alamo Chapter of Sigma XI. Finally, I was elected to the National Academy of Engineering. During this time I was included in a variety of Who's Who ranging from Who's Who in the South, Who Knows What and Where, to Who's Who in America, some dozen in all. Along the way I was selected as Most Distinguished Alumni of my *alma mater* and as an outstanding teacher by the national service society, Mortar Board.

Some honors will come to those who simply exist and rise through the ranks. Professorships are awarded for long service as well as for accomplishments as are university awards. National awards must be earned although it never hurts to have a friend in the organization who will speak a good word at the proper time.

CHAPTER 16

INVENTION

It has been said with some justification that necessity is the mother of invention. Sometimes, it is not so much a question of necessary as convenience. Scientists are not great inventors--the engineer holds that position by virtue of the fact that he has practical training in the applications of science that enable him to see a need for a particular device. The scientist, on the other hand, often sees a ready and perhaps unusual use for an invention once it has been developed.

It is also said that inventions are made for the returns. If so, I have failed to make proper connections. Most of the inventions with which I have been associated (Chapter 6) made money for the companies who produced a commercial product, but many of them were produced under government grant or contact and thus were not patentable.

I have always been a pragmatic and practical scientist and have seen needs in the laboratory. While at Emory University we were teaching medical students human physiology in large laboratories. One of the most pressing problems was measuring blood pressure in experimental animals. The usual method was to use a mercury column that was connected directly to a blood vessel. The top of the column contained a writing point that scratched on a soot coated revolving drum called a kymograph as the pressure moved up and down. The apparatus was cumbersome, easily breakable, and unreliable. The mercury was poisonous. Working with Harold Breding, the laboratory coordinator, we designed a capsule manometer. We constructed a thick plastic container about 5 inches in diameter and suspended within it a second container with very thin plastic walls. The outer container was connected to the blood vessel and the inner was filled with saline and connected to a recording device. As the blood pressure changed, the walls of the inner container expanded or contracted, and the change was recorded. With calibration against pressure measurements we had a readable, sturdy, sensitive and reliable measuring device. We built enough for an entire lab and used them until electronic recording came into practical use.

About the same time we were attempting to measure the respiration of cells of the pituitary gland to correlate it with synthesis of protein hormones. The device used was the Warburg respirator in which the tissue was placed in a small vessel connected to a manometer. Aa oxygen was used up and carbon dioxide produced the pressure in the vessel changed and the manometer recorded the change. There was a major problem in determining the volume of the vessel and the related changes in pressure that occurred. The usual method was to fill the vessel with mercury, weigh it and from that calculate the volume and then correlate the pressure reading on the manometer with pressure changes in the vessel as gases were consumed or released. This was a complicated and time consuming project. After some thought we made a device by which pressure could be changed in minute amounts and the volume calculated from it. This saved hours of weighing and calibrating. We published an article on the device, and made several of them for other Warburg apparatus users at our cost. Although the device might have been patentable, we did not do so because it was constructed on a government grant to solve a problem proposed in our research protocol.

In Pittsburgh the smog was everywhere. The snow was dark in color, the houses were yellow from acid fumes, and the steel mills emitted a reek that covered the whole city. When I arrived at the Industrial Hygiene Foundation at Mellon Institute all sorts of ideas were being studied to measure the extent of the smog. With the help of the machine shop I put together a device to solve the problem. A small vacuum pump drew air through a piece of filter paper at a constant rate. The filter paper was on a long roll that passed across the mouth of the inlet tube to the pump. At regular intervals a clockwork drew a fresh area of paper over the hole. At the end of the day we had a series of circles each representing the amount of air drawn through the filter paper in a given length of time. It was simple matter to read the density of the deposit that was a direct measure of the smog content of the air or to extract the material on the paper and analyze it for the actual contaminants. The device was used for several years in Pittsburgh to determine the times of day at which the greatest level of smog occurred and how the contamination changed.

Another invention was devised in Pittsburgh. We were working with environmentalists in Great Britain interested in how miners got "black lung disease" from inhaling rock and coal dust. The problem was also occurring in Appalachia. It was a debilitating disease with serious respiratory consequences. In order to investigate the problem it was necessary to know something about the deposition of dust in the lung When air enters the lung it passes through a variety of channels in the nose, trachea, bronchi, and lung and then is respired out through the same channels. It is possible that dust is deposited in all portions of the tract and that it is deposited on both inspiration and expiration. Since it is likely that dust deposited in the nose and trachea may not be as much of a hazard as that deposited in the lung itself, it is necessary to know the mechanics of deposition. We devised and built an apparatus to accomplish the task.

Several constraints entered into the design. It was necessary to control each breath to be exactly like the previous one in both rate and depth. It was essential that we be able to collect samples from various areas of the respiratory tract and repeat the collection at will. In order to simplify the experiment it was necessary to test various sizes of dust particles to see if they were deposited differently.

The apparatus consisted of a rotor that moved at a constant rate. The rotor had an input opening from the lung and the opening passed successively by a series of ports in the rotor housing which opened into electrostatic filters which collected the dust and then into bags where the volume could be measured. Thus as an individual breathed out successive volumes of air from the trachea, bronchi, and upper and lower sections of lung could be separately collected and analyzed for volume and dust content. The rate and depth of respiration was regulated by placing the subject in an old "iron lung" or mechanical respirator which we borrowed from the Harvard School of Public Health and the rotor was geared to the respirator so that repeated breaths from the same part of the respiratory tract could be collected at each rotation and accumulated for analysis.

The selection of the dust was critical. It had to be capable of being divided into various sizes of constant diameter, it had to be nontoxic, and easily handled. We looked at many materials and finally selected clay of the type that is used in pottery manufacture. The experiment was successful. The results are discussed in Chapter 5.

We also built a much reduced model which could be used for small animals such as the rat or rabbit. The device was used for testing materials we feared to use on humans.

My interest and areas of development in computers has been outlined in Chapter 2. The ideas were mine but the technical development must be credited largely to the Japanese.

Other inventions have come about because of research interest from which the invention sprang as an offshoot. This is certainly true of the use of o,p-DDD (Chapter 4) in the treatment of human adrenal carcinoma. We began with the idea of using the material to investigate the basic biochemistry of the adrenal cortex with little thought of medical applications.

Over the years there have been other minor developments. Several have been published in the *Review of Scientific Instruments* and other technical journals. Most devices were invented as a convenience in the laboratory without any thought of exploitation. In many cases the problems which were solved had already been solved but in a more difficult manner.

CHAPTER 17

PEREGRINATIONS AND PERSONALITIES

Scientists are inveterate travelers. They are willing to seize any excuse to attend a meeting, travel to another country, or present a seminar at another university. The symptoms are characteristic of a true disease which I have labeled Periomnitosis (moving disease). It is characterized by an intense desire to travel, the further away the better. The disease can be demonstrated in the young graduate student who seeks funds to present his first paper at a national meeting. It increases in intensity during the early and middle years of a scientific career and slowly dies as the scientist reaches towards retirement. Several factors exacerbate the problem. The availability of research funds (research grants contain funds for travel), reputation (a Nobel or other prize confers a great desire for travel), wide acquaintance in the field of science which permits trading of seminars and notoriety of the field of research (geneticists are asked to speak more often than zoologists).

The intense desire for travel is also increased by the opportunity for foreign travel. The sociologist who must investigate the slums of cities and picks a South American city to examine, the geologist who explores for oil in distant places, and the archeologist who digs in Turkey or Greece are all examples of carriers of the disease in its most malignant form.

The disease can be treated by the restriction of funds. It is clear that when grants are disapproved or travel funds restricted, the extent and time of travel decreases markedly. Another restriction is age. As travel becomes more difficult, the amount decreases.

The disease is encouraged by the national societies. Yearly meetings are held in places designed to attract visitors to Atlantic City, Anaheim, New York City and the like. The societies with the more affluent members, the medical specialities, for example, have meetings in exotic places, on cruise ships and in resorts. Most societies have international affiliations and dozens of international meetings are held each year, again in interesting places.

Another characteristics of the disease in the change in character of the afflicted individual. In the beginning attendance at meetings is faithful, papers are presented, and other speakers questioned. As the disease progresses, there is less attendance at the meeting and more conversation in the halls and bars. I recall the chairman of my department who went religiously to the FASEB meetings in Atlantic City each year and spent the entire week wrapped in a blanket sitting in the solaria of one of the large hotels. In another situation a high government official of my acquaintance in Washington left each Friday noon and returned on Sunday to spend the weekend with his mistress in New York at government expense.

Presentation of major papers usually takes place in a seminar and the ten minute paper at a large meeting is scorned unless it is absolutely necessary to assure travel. As aging progresses, the need for travel diminishes and as research grants decrease, funds must come from other sources. If a reputation has been acquired, presentation of keynote speeches, leading seminars and attending conferences may provide the necessary funds. If a major prize has been worn, the lecture circuit is available.

As a typical scientist, my travel serves as an illustration. During the early years I attended the national meeting of the FASEB and on occasion, the national meeting of the Endocrine Society. It was rare that I did not present a ten minute paper. Later I forsook the national meetings for the Laurentian Hormone Conference outside Montreal, Canada. It permitted wider discussion of topics, afforded a beautiful location, and allowed greater travel.

One of the attractions of NIH was the opportunity to travel. In my years with the Federal government, I managed to see all fifty states, a sizable area of South America, Asia, Europe, Australia, New Zealand, and parts of Africa. As time drew near for retirement, travel lessened. Now I travel largely on request to present a special topic, to serve as a keynote speaker, or lead a seminar.

NIH afforded tremendous opportunities to travel to meet people and to have people come to Washington for council meetings and consultations. I was very fortunate to be able to meet and talk with some of the most famous scientists and politicians of the day. One of the great collections of synopses of scientific articles is *Excerpta Medica* printed in Holland. The staff of *Excerpta* wanted to secure NIH support for their publications and I traveled to Amsterdam several times to discuss the matter. One of the Board members was the Nobel prize winner, Sir Peter Medawar. He was all business during the day but he liked to have a drink and see the town at night. We had several memorable evenings. The first Nobel Laureate of my acquaintance was Otto Myerhof. While at Rutgers I spent a summer at Bar Harbor Laboratories working on the blood pigments of lobster (which was a beautiful blue as contrasted to our red) and the Nobelist, Dr. Otto Myerhof was a summer resident. It was a small laboratory and we had time to chat about the Emden-Myerhof cycle that set a pace in biochemistry for the study of carbohydrate metabolism. Myerhof was old and fat with snow white hair and he spoke broken English.

The personalities of scientists ars interesting. I have been impressed with the change in Nobel laureates. Many stopped research (as most were aged this might be expected) to go on the lecture circuit. Others changed their field of research totally. This may result because the winner has spent too much time in a concentrated effort to achieve results in a relatively narrow field and simply wishes to rest and change. I remember Bob Glaser, an Atomic physicist who invented the Bubble Chamber. He turned to the design of a factory to produce microorganisms. We supported his work for a time as he tried to totally automate the process of growing plates of bacteria, reading and identifying them.

All scientists have a large ego. A prideful attitude must be present in order to assume that data is correctly determined, presented as a conclusion, and defended as a thesis against all comers. On the other hand, a scientist must be humble in order to appreciate the small measurements he/she is making on the complexities of nature. Most scientists are sure of their data. Even Einstein, the humblest of men, protested "God does not throw dice." when presented with the quantum statistical theory. In most successful scientists the intellect shows through. It was a pleasure to attend meetings of some of the committees the PSAC (President's Science Advisory Council) and listen to the conversation. I remember one meeting of the Psychology panel in Colorado Springs at the University of Colorado where George Miller, the great vision and color expert from Harvard, James Miller, the Rockefeller University psychologist interested in feedback, Herbert Simon from Carnegie Tech, a future Nobel prize winner, and others met for several days to discuss the future of training and research. I attended because NIH moneys would be used to fund the projects suggested. Conversation ranged over politics, psychology, future technology, and a variety of other topics.

The sons of famous scientists are a breed unto themselves. If they follow in the father's (or rarely, in the mothers) footsteps they begin at a disadvantage. Science is not like business where it is usual for the son to follow the father into it. Occasionally, the sibling also becomes great in science as did Eve Curie. But to emulate the older person is not usually possible. Many sons chose deliberately not to enter a scientific field. Others enter a different field from the father and attempt to become highlights there. I have known two such persons and they both worked with me in NIGMS. Trygve Tuve was the son the Merl Tuve, a great scientist with an international reputation at the National Bureau of Standards. Trygve was not able to emulate his father. He reached a level in the Institute and was unable to progress further. He died a young man under mysterious circumstances. The second person was Leo Von Euler. Leo came to America from Sweden, I think to avoid competition with his family. His father and grandfather were Nobelist and National Academy members. Leo was competent but not a great scientist and made his career as an administrator at NIH. He always appeared frustrated.

Still other scientists have used reputations gained in one field to become spokesman in another. The atomic scientist, Edward Teller, worked on the atomic bomb and then became an exponent of environmental research. Paul Erlich began as a biologist and wound up as an active proponent of environmental preservation. Linus Pauling, of course, set the record with a Nobel Prize in two entirely unrelated fields.

CURRICULUM VITAE

J. H. U. BROWN

GENERAL INFORMATION

Born in Nixon Texas, Married to Jessie Carolyn Schulz, 1943, no children. Served in U.S. Navy, World War II with honorable medical discharge. Grammar and High School in Texas. Diploma in 1936, Stockdale High School. Southwest Texas State University, 1936-1939, B.S. degree in Biology and Chemistry. Permanent teachers certificate. University of Texas, 1939-41. Graduate studies in Biophysics. Left because of war. Rutgers University, 1944-48. Ph.D. degree in Biochemistry. Numerous training institutes.

PROFESSIONAL RECORD

1989-. Professor Emeritus, University of Houston.

1980-1989. Professor of Biology, University of Houston; Professor of Business (Health Care Administration), Texas Womans University; Professor of Community Medicine (Adj.), Baylor College of Medicine; Professor of Public Administration, University of Texas (Adj.)

1978-1980. Associate Provost for Science and Research, University of Houston. Lecturer on Art, Continuing Education, University of Houston. Professor of Biology. Professor of Business (Health Care Administration), Texas Womans University; Professor of Community Medicine (Adj.), Baylor College of Medicine; Professor of Public Administration, University of Texas (Adj.)

1974-1978. Coordinator, Southwest Research Consortium. Professor of Physiology, University of Texas Medical Center, Professor of Environmental Sciences, University of Texas, Professor of Health Care Administration (Adj.) Trinity University. Lecturer, San Antonio Art Institute (Adj.), Coordinator of the Institutional Review Board

1974-78. Associate Deputy Administrator of the Health Services and Mental Health Administration (DHHS)

1960-1974. National Institutes of Health. 1974-Acting Director, National Institute of General Medical Sciences; 1972-74, Associate Director, NIGMS; 1968-72, Associate Director, Division of Research Resources; 1965-68, Assistant Director, DRR; 1962-65, Branch Chief, NIGMS; 1960-62, Executive Secretary, Physiology and Biomedical Engineering Study Sections. Served as acting Chief of Pharmcology/Toxicology Branch, Research Construction Branch, Clinical Research Centers and others. Lecturer in Physiology at Georgetown Medical School and George Washington University Medical School.

1952-1960. Associate and Full Professor, Acting Chairman of Department of Medical and Dental Physiology, Emory University School of Medicine.

1950-52. Assistant Professor of Physiology, University of North Carolina School of Medicine. Researcher, Oak Ridge Institute of Nuclear Studies.

1948-50. Director of Biological Research, Mellon Institute, Pittsburgh. Lecturer in Biology at University of Pittsburgh.

1944-50. Rutgers University. Fellow of Gerard Swope Fund; Lecturer in Physical Chemistry, Department of Chemistry, graduate student

1943-44. Lecturer in Physics, Southwest Texas State University.

1941-43. Instructor and Researcher, U.S. Air Force.

1939-41. Graduate student and Instructor in biology, University of Texas.

BIOGRAPHICAL REFERENCES AND HONORS

Biographical references include American Men of Science; Who's Who in Atoms; International Who's Who in Biomedical Engineering; Who's Who in West and Southwest; Who Knows What and Where; Who's Who in American Education; Who's Who in the World; International Dictionary of Biography; Who's Who in Technology; Men of Achievement and others.

Honors include Diplomate, National Academy of Engineering; Fellow, IEEE; Fellow, AAAS; Senior Member American Chemical Society; Senior Member, IRE; President, Founder, and Board Member, Biomedical Engineering Society; President, Founder, Board member, Alamo chapter of Sigma Xi; Founder, South Texas Print Society. Special Team award from NASA; Citation from NIH; Citation from *Emergency* Magazine; First Prize for documentary film, International film Festival; others.

Member of New York Academy of Sciences; Society for Experimental Biology and Medicine; American Physiological Society; National Council of Biology Editors; Endocrine Society; International Federation for Automatic Control; Alpha Chi(scholarship); Phi Lambda Upsilon (Chemistry honors); Pi Kappa Delta (forensic honors).

National Activities include Founder, President, Board Member, Biomedical

Engineering Society; Joint Committee on Engineering in Biology and Medicine; American Physiological Society committees; NASA/DHHS Joint Committee on Interagency Technology Transfer; White House Committee on Emergency Medical Care; PSAC, Committees on Psychology and Engineering; Editor for Academic Press, F.A. Davis; National Secretary, IEEE(BME); AAMC Education Committee.

BIBLIOGRAPHY-BOOKS

Brown, J.H.U.. *Physiology of Man in Space*. Academic Press, New York, 1960.

Brown, J.H.U. and S.B. Barker. *Basic Endocrinology*. F.A. Davis, Philadelphia. 1st Ed, 1962; 2nd Edition, 1966.

Brown, J.H.U., J.E. Jacobs, and L.E. Stark. *Biomedical Engineering*. F.A. Davis, Philadelphia, 1972

Brown J.H.U., and D.E. Gann. *Engineering Principles in Physiology.* Vol. 1 and Vol. 2. Academic Press, New York, 1973.

Brown, J.H.U., and J.F. Dickson, III. *Advances in Biomedical Egineering Vol* 2, 3, 4, 5, 6, 7. Academic Press, New York, 1972-1979.

Brown, J.H.U., *Integration and Control in Biological Processes*. Van Nostrand, New York, 1978

-----.*The Health Care Dilemma*. Human Sciences Press, New York, 1978.

----- *Politics and Health Care*. Ballinger, Cambridge, 1979.

-----.*Telecommunications in Health Care*. CRC Press, Boca Raton, 1982.

-----. *Management in Health Care Systems*. CRC Press, Boca Raton, 1983.

-----.*The High Cost of Healing*. Human Sciences Press, New York, NY, 1984

-----.*A Guide to Collecting Fine Prints*. Scarecrow Press, Metuchen, NJ, 1989

Brown, J.H.U., and J. Comola. *Educating for Excellence*. Greenwood Group, Westport, CT, 1991.

Brown, J.H.U., and J. Comola. *Improving Productivity in Health Care*. CRC Press, Boca Raton. 1988.

Brown, J.H.U., D. Mailman and R. Hazelwood. *A Laboratory Manual for Human Physiology.* Gorsuch Press, Tucson, 1985.

Brown, J.H.U., D. Mailman, and R. Hazelwood. *Animal Physiology* . Kendall Hunt Co. Dubuque, IO, 1989.

Brown, J.H.U. Editor for F.A. Davis for series:
Wunder, *Life Into Space*
Ferguson, *Blood and Body Functions*
Miller, *Energy Metabolism*
Brown, *Endocrinology*

INVITED CHAPTERS AND ARTICLES

Brown, J.H.U., and J.F.Dickson III. "Instruments in the Delivery of Health Services." Science 166: 334, 1969. Lead Article.
Brown, J.H.U.. "Telecommunications in Health Care." in Bourne, *Critical Rev. in Biomed. Eng.*, CRC Press, Boca Raton, 1980.
----."Is Medicine a Cottage Industry." National Forum. 83: 38, 1980.(Lead arttlcle)
----."Diagnostic Tests and Modern Therapy." in *Chem. and Modern Med.* Amer. Chem. Soc. Press, New York, 1976
----."Technology and Clinical Engineering" in Caceres, *Clinical Engineering,* Academic Press, New York, 1977.
----. "Parameters of Human Potential" in *Exploring Human Potential,* C.C.Thomas Springfield, 1966.
----. "Active Peptides of the Pituitary" in Croxatto, *Peptidos Activos,* Universitad Catolica, Santiago, Chile, 1969.
----. "Control of Energy Processes." in Ghista, *Applied Physiological Mechanics,* Harwood Publ., New York, 1980
----. Lead Articles in *Texas Medicine*: "Prevention", 75: 29, 1979; "Renal Dialysis", 75: 71, 1979; "Clinial Laboratories" 75: 59, 1979 ; "Emergency Care" 75: 63, 1979; "Reducing Costs of Care", 75:11, 1979 "Where Should You Have an Accident?",71: 69, 1975.
----. "Primates in Pharmacology." in Vogtborg, *Primates in Drug Evaluation*. Univ. of Texas Press, Austin, 1968.
----. "Futuristics." The National War College, Washington, 1972.
----. "A Personal History of Biomedical Engineering". Proc. 13th Symp. IEEE/BMES, IEEE Press. New York, 1990.

PUBLISHED PAPERS
(No abstracts or oral presentations are included; papers by invitation are indicated by asterisk)

Allison, J.B., J.A. Anderson, and J.H.U.Brown. "Evaluation of Protein in Hypoproteineinic Dogs" J. Nutr. 31: 327, 1948.
Seeley, R.D., J.B. Allison, J.H.U.Brown, and F.P. Ferguson. "Effects of Depletion and Repletion in Proteins on the Body Fluids of Adult Dogs." Proc. Soc. Exp. Bio. Med. 83: 214, 1948.
Brown, J.H.U., and J.B. Allison. "Effects of Excess Dietary Methionine and Arginine on Adult Rats." Proc. Soc. Exp. Biol.Med. 71: 812, 1949.
*Brown, J.H.U., K.M.Cook, G. Ney and T.F. Hatch. "Influence of Particle Size on the Retention of Airborne Particulate Matter in the Human Lung." Amer. J. Public Health. 40: 450, 1950.
Brown, J.H.U., and P.F. Gross. "The Mechanism of the Anaphylactoid Phenom-

ena." J. Med. Sci. 221: 45, 1951
Gross, P.F., J.H.U. Brown and T.F. Hatch. "Production of Endogenous Lipid Pneumonia." Amer. J. Path. 27: 691, 1951.
Brown, J.H.U., K.M.Cook, G. Ney and T.F. Hatch. "An Apparatus for Partitioning Expired Air." Amer. J. Clin Path. 22: 351, 1952.
Gross, P.F., and J.H.U. Brown. "Endogenous Lipid Pneumoniotic Deposits." Amer J. Path. 22: 351, 1952.
Gross, P.F., and J.H.U.Brown. "Mobility of Pneumoniotic Deposits." Amer. J. Path. 23: 352, 1952.
Brown, J.H.U.. "Metabolism of Methionine.". J. E. M. Scientific Society, 68: 11 1952.
----. "Action of an Adrenocorticolytic Agent on the Adrenal Cortex of the Rat." Proc. Soc. Exp. Biol. Med. 88: 59, 1953.
----. "The Adrenocorticolytic drug DDD." U.S. Air Force Bulletin # 1916, 1953
Brown, J.H.U., K.M.Cook, L.E. Palm and T.F. Hatch. "An Apparatus for the Partitioning of Expired Air in Small Animals." J.Rev. Sci. Instr. 25: 657, 1954.
Brown, J.H.U.. "An Improvment in the Reddy Method for the Determination of Steroids in Urine." Metabolism 4: 295, 1955.
Brown, J.H.U., and J. H. Bain. "A Simple Warburg Calibrator." J. Lab and Clin. Med. 45: 822, 1955
Brown, J.H.U., J. Griffin and R.B. Smith,III. "Excretion of Urinary 17-hydroxycorticoids in Dogs Fed DDD." Metabolism.4: 542, 1955.
Brown, J.H.U.. "Effective Lung Ventilation in the Artificial Respirator." J. Med. Assn. Ga. 44: 229, 1955.
Gross, P.F., J.H.U. Brown, R.P. Scric, N.L. Butler and M. Westrick. "The Toxicity of Calcium Halophosphors-I." Arch. of Inter. Med. 11: 479,1955
Gross, P.F., J.H.U.Brown, R.P. Scric, N.L.Butler and M. Westrick. "The Toxicity of Calcium Halophosphors-II." Arch. Inter. Med 11:485, 1955.
Brown, J.H.U., J. Griffin, R.B. Smith and A. Anason. "Excretion of 17-hydroxy corticoids in Dogs Fed DDD." Metabolism 4: 522, 1955.
Brown, J.H.U.. "Determination of Steroids in Urine." J. Med. Assn. Ga. 44: 529, 1955.
*----. "Review of Physiology." Nursing Outlook 4: 303, 1956.
Hess, M., and J.H.U. Brown. "Separation of the Trophic Hormones of the Anterior Pituitary by Ultracentrifugation." Anatomical Record. 124: 303, 1956.
Jacobs, J., and J.H.U. Brown. "Kidney Function in the Adult Rat." Amer. J. Physiol. 186: 172, 1956.
Brown, J.H.U., R.B.Smith, J.R. Griffin and J. Jacobs. "Influence of Cortisone and the Action of an Adrenocorticolytic Drug." Endoc. 61: 106, 1955
Cueto, C., and J.H.U.Brown.. " A New Stain for Dog Eosinophils." Stain Tech. 33: 95, 1958

----. "The Chemical Fractionation of an Adrenocorticolytic Drug." Endoc. 62: 326 1958
----. "Biological Studies of an Adrenocorticolytic Drug." Endoc. 62: 334, 1958
LaBella, F., and J.H.U.Brown. "Distribution of Hydrolases Among Anterior Pituitary Cell Fractions." J. Biophys. Biochem Cytol. 4: 833, 1958.
----. "Cell Fractionation of Anterior Pituitary Glands" J. Biophys Biochem Cytol. 5: 17, 1958.
Brown, J.H.U., and A. Anason. "*in vitro* Metabolism of Substituted Steroids by Rats Liver." Endoc. 62: 103, 1958.
Brown, J.H.U., and C.H. Brown. " Action of an Adrenocorticolytic Drug in Man" Lancet, June 6, 1959, p. 1208.
Brown, J.H.U., and F. Ulvedal. "Studies on Ultracentrifuge Fractions of Anterior Pituitary." Endocrinology 66: 182, 1960.
Brown, J.H.U., F .Labella and F. Ulvedal. "Studies on the Intracellular Localization of Anterior Pituitary Hormones." Endoc.66: 1, 1960.
Brown, J.H.U., and F. Ulvedal. Studies on the Ultracentrifuge Fractions of the Anterior Pituitary." Endoc. 66: 175, 1960.
*Brown, J.H.U.. "Specific Inhibitors of the Adrenal Cortex." Nature 187: 985, 1960
Brown, J.H.U., and A. Petkas. "Behavior of Kidney Mitochrondia Under the Influence of ADH." Endocrinology 69: 182, 1961.
*Brown, J.H.U.. "The Engineer and the Life Sciences." Proc. IRE 50: 1917, 1962.
*----. "The Physiologist and the Engineer." Proc. Instru. Soc. Amer. 8: 137, 1962.
----." Training of a Physiologist." The Physiologist. 5: 248, 1962.
Brown, J.H.U., C.P. Brower, H.B. Pahl, T. Tuve and L. Hellman. "Ideas or Instruments?" J. Med. Ed. 37: 12, 1962.
Brown, J.H.U.. "Engineers and the Medical Sciences." Abbotempo 1: 38, 1963
----. "The Physician and the Engineer." J.A.M.A. 82: 730, 1962.
----. "General Clinical Research Centers." J.Med.Ed. 40: 1482, 1965.
Stone, F.L., and J.H.U.Brown. "Technology and Medicine." Biosci. 15 : 71, 1965
Stone, F.L., and J.H.U. Brown. "Manpower in the Basic Medical Sciences." J. Med. Ed. 42: 500, 1967.
Brown, J.H.U.. "New Horizons in Biomedical Engineering." Biosci. 17: 391, 1967.
----. "The Science Administrator." J. Med.Ed. 43: 33, 1967.
Brown, J.H.U., and F.L. Stome. "A Program in Drug Research" J. Amer Pharm. Assn. 58: 439, 1968.
Brown, J.H.U, and R. Melville. "Technology and Medicine." Lab. Mgte.7: 25, 1969.

Brown, J.H.U.. "Clinical Pharmacology." J. Med. Ed. 44: 1351, 1969.
Hening, A., and J.H.U.Brown. "Mutagenesis." Science 163: 1355, 1969.
Brown, J.H.U.. "The Behavorial Sciences in the Medical School." Science 163: 964, 1969.
----. "Parapetosis-A New Disease." Physiologist. 14: 2, 1971.
*----."The Biomedical Engineer in the U.S." Biomed. Eng.(UK) 6: 405, 1971.
Brown, J.H.U., and D.J. Lowell. "Standardization and the Delivery of Health Services." Proc. IEEE 20: 218, 1973.
*Brown, J.H.U.."Technology in Clinical Medicine." Clin. Med. 1: 4, 1973.
----. "Technology in Modern Medicine." Trans. IEEE(BME) 20: 213, 1973.
----. "Biomedical Eng20: 95, 1975.
----. "Developing a Health Care System." Ann. Biomed. Eng. 3: 3, 1975.
*----."Problems of R and D in Health Services." Ann. Biomed. Eng. 3: 231, 1975.
*----."The Crises Emergency System." Emerg. Med. Ser. 6: 58, 1977.
----. "Medical Devices as Medical Problems." Biomed. Eng. 11: 337, 1976.
----. "Control of Health Care Quality." Trans. IEEE(SMC) 6: 752, 1976.
----. "Communications and the Delivery of Health Care.". Trans IEEE (SMC) 7: 60, 1977.
----. "The Consumer and the Health Care Feedback Loop." Trans IEEE (BME) 7: 465, 1977.
----. "The Research Consortium." J. Res. Adm. 9: 8, 1978.
Brown, J.H.U., L. Schonfeld and P. Allen. "The Management of An Institutional Review Board." J. Res. Adm. 10: 5, 1978.
Brown, J.H.U.. "Functions of an Institutional Review Board." Fed Proc. 38: 2049, 1979.
Brown, J.H.U., L. Schonfeld, and P. Allen. "Costs of an Institutional Review Board." J.Med. Ed. 54: 234, 1979.
Brown, J.H.U.. "Philosophy of an Institutional Review Board." J. Med. Ed. 55: 68, 1980.
Brown, J.H.U.. "Research Consortia." Res. Mgte. 24: 38, 1981.
----. "Art at Auction." Art Happenings, 5: 4, 1980.
----. "New Concepts in Health Care Delivery." Texas Med. 78: 25, 1982.
----. "Medical Schools in Crisis." J. Eval in Health Prof. 11: 149, 1988.
Brown, J.H.U., and C. Vallbona. "The Laser Card as a Medical Record System." in *Optical Information Systems*. Meckler, New York, 1987.
----. "A New Patient Record System." Proc. Eighth Conf. on Bioeng. New York, 1986. p. 110.
----."A New Patient Record System." Proceed. Conference on Medical Systems, Boston, 1987, p. 116.
----. "The Laser Card in Ambulatory Care." Proc. Conf. Community Med.Soc. Community Med. Press, Kansas City, 1988.
----."New Forms of Patient Records." Proc. Conf. on Computerization of

*----. "The Optical Card in Medical Practice." Computers in Biol. and Med. 19: 375, 1989.

*Brown, J.H.U., C. Vallbona and J. Shoda. "Evaluation of the Optical Card Record System" in Roth Symp. on Computer Applications in Medical Care, IEEE Press, Washington, 1989.

*Brown, J.H.U.. "Bone Growth and Dissolution." in Magill's Survey of the Sciences, Salem Press, Pasadena, 1991.

*----. "Telecommunications in Health Sciences." Proc. Intern. Cong.on Telecommunications, University of Calgary, Canada, 1990.

Shoda, J., C.Vallbona, J.H.U. Brown and J. Albin." Evaluation of Optical Memory Card System for Patient Records." Proc. Intern, Optical Soc., Vol .1348, 1990, p.536.

Browm, J.H.U.., "The Research Consortium." IEEE Trans in Comm..June 1991.

----. "Evaluation of a Laser Card Patient Record System." Comp in Biol and Med 13: 651.1991

*Brown, J.H.U., C. Vallbona and J. Shoda. "Pilot Studies on the Laser Card" in *Smart Cards*, AEA, Santa Clara CA., 1991.